River Basin Morphology

with bamboo groves to meet the demand of rural house - building material and to check soil erosion and destruction of the river bank during floods in rainy season (Plate 6.6).

6.2.[illegible] BUILT UP LAND

Built up land covers only 0.78 sq km or 0.06 per cent of the total basin area. This shows that this drainage basin is least developed with very little urbanisation and poor economic growth.

6.3 LANDUSE PLANNING AND WATERSHED MANAGEMENT

In the following paragraphs an attempt has been made to analyse the existing landform characteristics and present land use pattern. The various limitations and impediments in the utilisation of land resources are all related to physiography, rainfall characteristics and cultural factors. The watershed approach can be integrated with programmes like forestry, soil conservation, rural and community development and farming systems (NIRD, 1989). It is a well known fact that land and water interact to produce food, fodder, firewood, timber and raw materials for the people, their domestic animals and agrobased industries. Therefore a well planned landuse strategy is essential for obtaining the most productive interaction between these resources.

The following suggestions have been made for proper landuse management in the Iril river basin.

6.3.1 PLANNING STRATEGY

It is necessary to solve the problems that arise while planning the landuse in the Iril watershed and suggest plans to minimize the misuse of land resources. Singh (1971) proposed that "the landuse should be re-oriented with a flexible and dynamic plan aimed at providing maximum utility for every acre of land". However in this context land fragmentation is one of the major problems in the region. Land use on private land cannot be influenced by mere legislation. There are other drawbacks in planning, viz. lack of communication facility, financial support, proper understanding etc. Furthermore, though [illegible] land is recommended for [illegible] production, it cannot materialise due to lack of financial support to cater consumption expenses during intervening period etc. Therefore, the problem of landuse changes should be tackled

River Basin Morphology

Hijam Ibeyaima Devi

RAJESH PUBLICATIONS
New Delhi - 110 002

First Published in India : 2000
ISBN : 81-85891-34-6

Published by :
M.L. GUPTA
RAJESH PUBLICATIONS
1, Ansari Road, Darya Ganj,
New Delhi - 110 002

Type Setter :
Anita Gupta
Sanjay Enterprises, New Delhi-110 002

Printed at :
Graphic Print
New Delhi-110 002.

PREFACE

The geomorphological studies have wide applications in planning and management of agricultural and horticultural landuses, crop selection with reference to soil quality and irrigation facility, development of transport and communication, industry and settlement etc. in addition to its importance, in its own right, as a well defined branch of earth sciences.

The present study of geomorphology of the Iril river basin is a modest attempt to explain the actions of running water affecting the features of the landscape termed popularly as fluvial geomorphology. The analyses of the basin characteristics reveal various litho - structures and processes of landform evolution, both natural and human induced. The basin morphometry hints at a dynamic combination of various agents *viz.*, neotectonism, intensive monsoon rain under a subtropical climatic regime and unscientific landuse practices, acting in concert to cause the evolution of a variegated polycylic landscape in the Iril basin.

In view of the rapid progress of land degradation due to erosion of topsoil from the denuded hill slopes of the catchment, a number of landuse changes have been suggested for the Iril river basin. The size and features of the basin are ideal for its adoption as a model watershed for spatial integration of various developmental components.

I express my deep sense of gratitude to Professor R.P.Singh, ex-Head, Department of Earth Sciences, Manipur University for his guidance, supervision and unceasing encouragements during the course of my investigation. I thank him also for providing me the facilities available in the Department during my study.

I am indebted to Prof. R. Gauntia, Department of Geography, Patna University; Prof. H.R. Betal and Prof. S.C. Mukhopadhyay, Department of Geography, Calcutta University ; Dr. P.P. Rao and

Dr. P.N. Rao, GSI Photogeological Training Centre, Hyderabad and Dr. Arun Kumar and Dr. R.A.S. Kushwaha, Department of Earth Sciences, Manipur University for their valuable suggestions and encouragements.

The valuable suggestions, encouragements and material helps I received from the following senior officers, colleagues and friends are gratefully acknowledged : Shri Dhiraj Yumnam, Addl. Director of Industries, Shri Harsh Gupta, Dy. Director, Shri Kh, Dilip Singh, Geologist, Smt. Bimolata Devi Chingakham, Geologist, and Shri P.P. Dev, Asst. Geologist, Geology and Mining Division, Department of Industries, Govt. of Manipur; Shri Dinachandra Singh, A.E.O. Remote Sensing Cell, DSTE, Govt. of Manipur ; Shri Sharat Singh, Dy. Director of Economics and Statistics, Govt. of Manipur; Mrs. Joya Ghosh, Sr. Lecturer, Presidency College, Shri H. Akendra Sharma, Lecturer G.P.Women's College and Dr. Ksh. Bimola Devi of Mayai Lambi College.

I thankfully acknowledge the help extended to me by Shri I. Tomba Singh, Draftsman, Industries Department, Manipur in preparation and final drawing of maps and diagrams used in the book.

I sincerely acknowledge my indebtedness to the authors of various books and articles listed in the Bibliography whose learned works have immensely helped me in completing this study.

Lastly, I express my indebtedness to my husband Dr. R.K.Shyamananda Singh for his countinued help on both the academic and domestic fronts without which it would have been extremely difficult for me to complete this work.

Beikunthanath Mandir,
Palace Compound,
Imphal - 795 001

Hijam Ibeyaima Devi

CONTENTS

LIST OF TABLES

LIST OF FIGURES

LIST OF PLATES

Barail group of rocks are predominantly arenaceous in character with thick beds of sandstone, shale and siltstone mostly occurring at the core of the major synclines with gradational contact and marks of coarse gritty sandstone. The rock types, found in Disang group, are somewhat argillaceous in character with predominance of shale and siltstone. Siltstone beds are relatively not common. The siltstone found in the basin is generally soft and friable in nature. Major part of the basin is covered by the rocks of Disang formation.

This area has a complex history of geological evolution which is indicated by multiple folding, faulting and shearing. Three different phases of folding are observed in the Iril basin area. The first generation of folding was in the form of localised tight isoclinal to open type of folds with NE-SW trending sub-vertical areal planes and low to moderate plunges. The second generation of folds were in the form of regional anticlines and synclines affecting the formations and showing various types of folding from open asymetrical to tight symmetrical folds. The third generation of folds came in the form of broad warping, trending almost E-W, characterised by prominent and persistent upstanding feature of sandstone and siltstone in Mungching and the areas north, NE, NW and east of it. Many of the major strike faults have restricted the domain of different phases of folding and have also controlled the topography and drainage of the basin area. The middle part of the Iril river itself is controlled by major fault *i.e.* the Iril fault. Some of its tributaries too flow along the fault lines.

The type of index microfossils found in the basin also support the presence of Barail and Disang group of rocks. Landforms such as cliffs, alluvial fans, terraces and erosional surfaces are the characteristic features of the basin which are encountered in course of its downward movement. The valley portion is mostly covered by fertile alluvium, which contains deep silts deposited by the Iril and its tributaries.

Climatically this basin area enjoys a humid to subtropical monsoon climate. However, this climate largely prevails in the relief region upto 1800m, and the higher regions enjoy more or less a temperate climate. Annual rainfall is sometime as low as 837 mm and sometime as high as 2047 mm and temperature ranges between 1.5° and 37°C.

1

INTRODUCTION

The present treatise is concerned with the study on the complex fluvial geomorphological aspects of Iril basin within the Manipur basin. River Manipur is a tributary of the Myittha river of the Chindwin-Irrawaddy drainage basin of Myanmar (Burma). The Iril river originates from the south - eastern slope of the Hougdu Ching range in Ukhrul district of Manipur at a height of 2473 m. above the mean sea level (MSL) and flows over the unstable formation of the hilly tract. It is a tributary of Manipur river, which is called Imphal river in its upper part before its confluence with Khuga river. The area covered by the Iril and its distinct tributaries forms a sub - basin of the Manipur drainage basin. In the present study, the Iril sub - basin has been termed as the Iril basin accepting the drainage area as a distinct geomorphic unit. Further, Iril basin is marked by the development of a number of fluvial landforms including micro landforms of the erosional and depositional origins, which present many interesting geomorphological features related to the evolution of a variegated polycyclic landscape. Hence, the term Iril basin has been used instead of Iril sub - basin in the present geomorphological study.

1.1 LOCATION

An area of 1,285 sq km, covering the north eastern portion of the upper Manipur basin in the districts of Ukhrul, Senapati, Imphal East and Thoubal, forms the Iril river basin, which lies between the latitudes $24^{0}40'$ N-$25^{0}25'$ N and the longitudes $93^{0}55'$ E-$94^{0}20'$ E (Fig. 1.1). The total length of the Iril river from its head to the point of its confluence with the Imphal river is 144.5 km. Though the Iril basin covers a small area, it is characterised by a complex and intricate drainage system. The typical features exhibited are believed to have been produced by the interactions of the underlying folded structure, varied rocks and tectonics on the one hand, and varied geomorphological processes specially

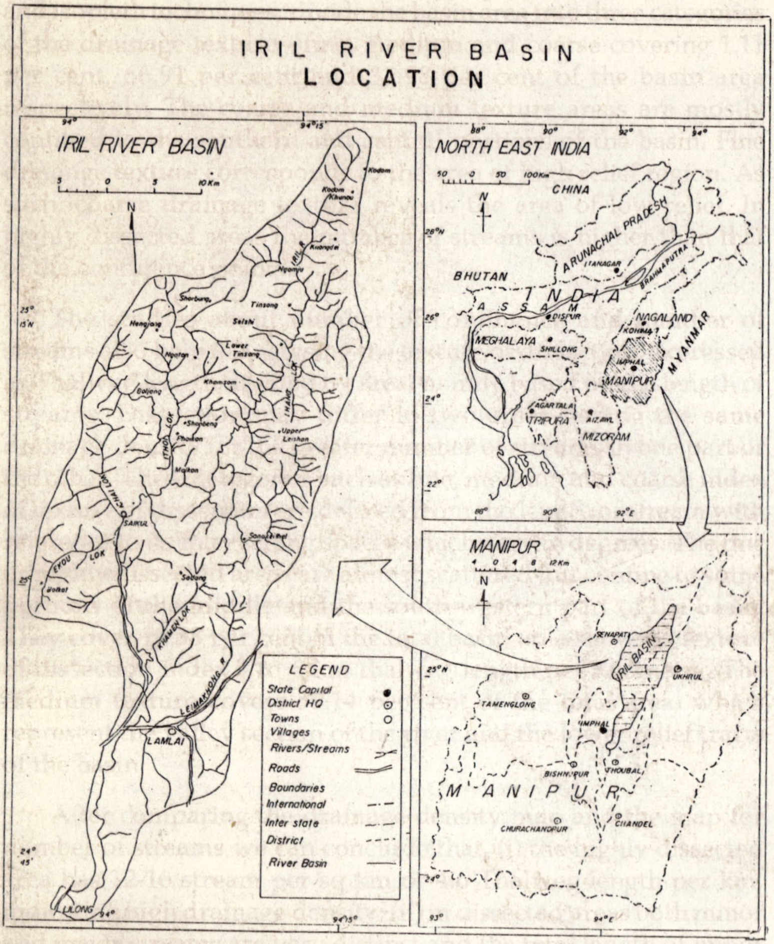
IRIL RIVER BASIN
LOCATION
IRIL RIVER BASIN
NORTH EAST INDIA
CHINA
ARUNACHAL PRADESH
BHUTAN
INDIA
ASSAM
NAGALAND
MEGHALAYA
MANIPUR
MYANMAR
TRIPURA
MIZORAM
MANIPUR
LEGEND
State Capital
Distict HQ
Towns
Villages
Rivers/Streams
Roads
Boundaries
International
Inter state
District
River Basin
LAMLAI
LILONG
SENAPATI
UKHRUL
TAMENGLONG
IMPHAL
BISHNUPUR
THOUBAL
CHURACHANDPUR
CHANDEL

Fig. 1.1

the fluvial processes on the other.

On the west of the source of the Iril river is the water divide beyond which water is drained by the tributaries of the Barak river, which flows along the folds of the western hills of Manipur, forming a tributary of the Ganga - Brahmaputra drainage system, which finally emits into the Bay of Bengal. The long ranges of hills act as the watershed between the Imphal and Barak rivers. Beyond the water divide on the upper eastern part of the Iril river are the small tributaries such as Pei Lok that flows towards Nagaland, the neighbouring state and the Laini Lok which pools its water into Bazaru or Knobari having a big bend towards the western side. The lower parts, beyond the eastern water divide of the Iril, are drained by the small tributary streamlets of Imphal river. From a point between the heads of the Iril and the Barak rivers originate the Imphal river, one of the main rivers of Manipur. It flows southwards and receives the water of the Iril river.

1.2 OBJECTIVES OF THE STUDY

The present study of a small river basin in the north eastern part of Manipur state is an attempt in fluvial geomorphological analysis. The existing geological and geomorphological informations on Manipur are not adequate for evolving a [illegible] and acceptable model for the geomorphic evolution of the catchment of the Iril river. Therefore, the present study aims to evaluate :

a. the stratigraphic set up, tectonic and the structural conditions,

b. the river basin morphometry,

c. the variations and inter - relationships of the morphometric parameters of third order drainage basins,

d. the landscape pattern,

e. the interpretative account of the development of fluvial landforms and drainage pattern as influenced by the underlying structure and tectonics under the multiple cycles of fluvial erosion,

f. the magnitude spectrum of the landforms correlating the drainage patterns and the structures, and

g. the characteristic landuse pattern of the Iril river basin.

1.3 METHODOLOGY

The study of characteristics of landforms depends on appropriate map scale and kind of study, and needs clear understanding of the existing structural parameters. The analysis of geomorphological features involves correct interpretation of the information collected from maps and field works. The topographical maps used in the present analysis are Survey of India topographical sheets 83K/3, 4, 7 and 8; 83G/16; 83L/1; 83H/13 and 14 on the scale 1:50,000. The maps reveal the characteristic features of the study area. Topographical maps on this scale show contours at 20 meters interval which omit many minor landforms. Again geological maps of this area, except those of some parts of it, drawn by the Geological Survey of India (GSI) as a part of their field work, are not available. This creates another difficulty for the study of the structure and geology of this basin area. Other data collected for the analysis such as those on rainfall, temperature, humidity, stream discharge and reports on geological mapping are from such sources as the Meteorological Centre, at Guwahati Airport and Irrigation and Flood Control Department, Directorate of Economics and Statistics, Geology and Mining Division of Commerce and Industries Department and Department of Science, Technology and Environment, Government of Manipur.

The analysis of the structural conditions, stratigraphic set up, lithology, geological features, such as folds, faults, joints etc. is based on the geological maps drawn by the GSI. The drainage map of the study area has been drawn with the help of 1:50,000 topographical sheets published by the Survey of India.

The number of confluences, number of streams, length of streams etc. have been counted on the drainage map and shown in separate maps using the choropleth and isopleth techniques.

The isopleth maps show the pattern of the density categories while the choropleth maps depict such categories within bounds of the squares measuring one square kilometer each. The stream line surface map has also been prepared, which indicates the amount of denudation of the land surface. The stream line surface has been obtained by plotting intersections of actual contours and main stream.

1.3.1 MORPHOMETRIC PARAMETERS

Morphometry deals with the measurement and mathematical analysis of the configuration of the earth's surface and of the shape and dimensions of its landforms (Clark, 1966). Drainage or fluvial morphometry, on the other hand, denotes the measurement of geometrical properties of the land surface of a fluvial erosion system. The basic form elements of a fluvial erosion landscape include linear properties such as number, length and arrangement of streams, areal properties like surface area and shape of drainage basins, and relief properties such as relief (magnitude of vertical dimension of landforms) and gradients (slopes) of ground surfaces and stream channels (Strahler, 1964).

According to Horton's Law (1945) and Strahler's system (1952) of stream order and number, the area has been divided into units of hierarchical order of different magnitudes.

a. Stream Order and Number of Streams

The number of streams gradually decreases with the increase in stream order. The first step in the drainage basin analysis is the determination of stream order. Stream order is a measure of the position of a stream in the hierarchy of tributaries (Leopold *et al.*, 1964). It offers a quantitative basis for the comparison of the degree of development in the drainage network of comparable size (Horton, 1945). Further, it is obvious that the number of streams is dependent on the type of lithology also.

b. Stream Lengths

The total lengths of the stream of different hierarchical orders and their mean values are of great importance in the study of drainage basin. Stream length is a result either of the number of stream segment or of the presence of larger segments. With every increase in stream order, its segment length tends to increase. The average length of the stream of each order in a drainage basin tends closely to appropriate a direct geometric series in which the first term is the average length of the streams of first order and constant factor is order ratio (Horton, 1945).

c. Length of Basin

Length of basin suggest the shape of the basin and can also show relative elongation between one basin and another.

d. Basin Area

Basin area and the length of streams are significantly correlated. The highest value of 3rd order basin area and the total length explains a significant linear relationship also. If length of stream is plotted against basin area, it tends to follow a distinct trend. It supports the hypothesis that the greater the value of the area, the higher will be the associated value of the total stream.

e. Basin Relief

Relief of a basin may be defined as the difference in altitude between the highest and the lowest points. Relief has a close relation with junction angles. Junction angle varies inversely with relief (Lubowe, 1964).

f. Bifurcation Ratio

Horton (1945) established a geometrical relationship among stream length, stream number and stream order. According to him, the number of streams of different orders in a given drainage basin tend closely to approximate an inverse geometric series in which the first term is unity and the ratio is the bifurcation ratio. Thus, bifurcation ratio is the slope of the line relating number of streams to stream order (Leopold *et al*, 1964).

The bifurcation ratio (R_b) has been worked out as a ratio between the number of streams of a given order (u) and the number (N_u) of the streams of the next higher order. The following formula has been used to work out the bifurcation ratio :

$$R_b = \frac{N_u}{N_u + 1}$$

Where, N = Number of streams

u = Stream order

g. Drainage Density and Stream Frequency

Drainage density is the length of stream per unit area. It is expressed as follows :

$$D_d = \frac{\Sigma L}{A_d}$$

Where, D_d = Drainage density

L = Length of streams

A_d = Area of drainage basin

Drainage frequency or stream frequency is obtained by using the following formula :

$$F = \frac{N}{A}$$

Where, F = Frequency of streams

N = Total number of streams

A = Area of the drainage basin

h. Drainage Texture

The drainage texture means "the relative spacing of drainage lines" (Thornbury, 1957). It includes both the drainage density and the stream frequency (Horton, 1945). It is directly related with drainage density and is influenced by climate, infilteration capacity, mean available relief, geological structure etc.

i. Basin Shape

The stream discharge characteristics of a drainage basin can be affected by its shape. The shape of a drainage basin can generally be expressed by :

i. Form Factor (F_f)

$$F_f = \frac{W_b}{L_b} = \frac{A}{{L_b}^2} \qquad [\because A = W_b \times L_b]$$

Where,

W_b = axial width of basin

L_b = axial length of basin

i.e. the distance from the measuring point to the most remote point on the basin.

ii. Compactness coefficient (C_c)

$$C_c = \frac{P_b}{2\sqrt{\pi A}} = \frac{P_b}{2\sqrt{\pi\left(\pi R^2\right)}} = \frac{P_b}{2\,\pi R}$$

Where Radius R of an equivalent circular area is given by Area = πR^2.

Therefore, the value of R is found out.

iii. Elongation Ratio (E_r) is defined as the ratio of the diameter of a circle of the same area the basin to the maximum basin length (Schumm, 1956). This ratio can be expressed by :

$$E_r = \frac{2R}{L_b}$$

Where R = Radius of an equivalent circular area which is given by Area = πR^2 and thus the value of R is found out.

L_b = Length of the thalweg (the distance from the measuring point to the most remote point on the basin).

iv. Miller (1953) used a dimensionless "Circularity ratio (Cr)" defined as the ratio of the basin area to the area of a circle having the same perimeter as the basin. This ratio can be expressed by :

$$C_r = \frac{A}{R^2}$$ Where A = area of the basin

Radius R of a circle of an equivalent perimeter as the basin is given by,

$2\pi R$ = Perimeter of the basin and thus the value of R is found out.

1.3.2 RELIEF CHARACTERISTICS

Absolute and the relative relief, various elevation groups, landscape profiles, slope categories and intensity of relief form important features of the drainage basin.

a. Area Height Curve

The area height curve (or diagram) represents the absolute correlative areas of land between adjacent contours (Clark, 1966). It is helpful in determining the surfaces of a region. Hypsometric curve or hypsographic curve or absolute hypsometric curve is the measurement of the interrelationship of area and altitude (King, 1966). The altimetric frequency histogram deals with the frequency of occurrences of different elevation groups in a region.

b. Landscape Profiles

Superimposed, composite and projected profiles of the basin landscape have also been drawn. The superimposed profile brings out surface with different slopes, corresponding to periods of peneplanation and reveals significant surface forms. The projected profiles show accordant summit levels and the composite profile shows the surface ruggedness and the general skyline of the landscape.

c. Slope

The word slope or gradient reflects many factors such as intensity and relative relief, texture and density of the drainage, tectonic structure of the particular region, climate and geology. Maps of relative relief, intensity of relief and slope category of the area have been prepared by using the methods of Smith (1935), Raisz and Henry (1937) and Wentworth (1930).

Intensity of relief has been calculated by finding out the average relief per unit area, *i.e.* per sq. km. According to Smith (1935) relative relief has been calculated by determining the difference in height between the highest contour point and the lowest points in each sq km area. The absolute relief region has also been calculated as the highest point in each sq km area.

Slope categories were determined by the average slope determination method (Raisz and Henry, 1937). Accordingly the total basin area has been divided into small regions, within each of which the contour lines have the same spacing *i.e.* the same number of contour lines per sq km. Another method of average slope determination, formulated by Wentworth (1930) was also followed. Under his scheme, the whole basin was divided into several rectangles for the sake of working convenience. The contour map of the basin area was put under an east - west and north - south transparent grid in metric scale and the total number of contour crossings along each grid line was measured and thus the average number of contour crossings per kilometre was determined. The procedure was repeated using an oblique grid over the same area and the readings of the two operations were averaged. The results were then converted into degrees and the average slope map was prepared based on the data worked out by applying the following formula :

$$\text{Slope in } \tan\theta = \frac{\text{Total no. of contour crossings} \times \text{contour interval}}{\text{Total grid length} \times \text{constant}}$$

$$\text{Average Slope} = \frac{\text{Total no. of contour crossings} \times 100}{\text{Total grid length} \times 636.6}$$

Where the constant 636.6 is adopted for metric system derived as 0.6366 x 1000 (No. of metre in 1 km.) in place of 3361 (derived as 0.636 x 5, 280 *i.e.* No. of feet in mile) in FPS system. The value 0.6366 is the mean of all possible values of sin θ where θ is the horizontal angle between the contour and grid line.

1.4 REVIEW OF LITERATURE

The history of geomorphology can be traced back to the days of Greek, Roman and Arab philosophers and historians. However, modern geomorphology is the result of the works of the eighteenth and nineteenth century geologists and hydrologists such as Lamark and Du Buat (Kumar, 1979). However, the man who laid the foundation of modern geomorphology was the first great fluvialist, James Hutton (1795). The eighteenth century Neptunists propounded a very crude theory of erosion, one invoking great floods. But, the nineteenth century Plutonist's theory, according to which valleys originated as upper fissures during mountain upliftments and were only modified in detail by water and the forces of weathering and degradation, gained the upper hand (Hettner, 1928). However, both the theories including that of Hutton and Hettner were rejected by Lyell (1872), who took the features as products of marine action. In England, Greenwood (1857) and Beeta Jukes (1862) put forward the view that valleys had been carved out by the rivers that occupy them by the process which continued over thousands of years (Hettner, 1928).

Davis (1915) was largely responsible for classifying valleys. During the 1920s and 1930s, Walther Penck (1924) and his followers staged a minor revolt against some of Davis' idea and found some adherents (Thornbury, 1985). In the mid 1870s, Powell (1875) introduced the ideas of consequent valleys and later Davis added more nomenclatures which are rarely used nowadays. Powell attempted to classify landform with the help of geologic structures and devoted much attention to stream erosion. No one had ventured before Powell to explain the sculpturing of the land by impact of rain and stream beyond what today we would call late maturity in the geomorphic cycle. Powell recognised that the process of erosion operating undisturbed upon the land, would eventually reduce it to a lowland somewhat above the sea level (Thornbury, 1985). Research on the relationship between geomorphic processes and resultant landforms was revived by Reiche and Keller for evaluation of the process of weathering; Hjulstrem and Sundborg for fluvial action; Krumbein Zenkovich and King for marine agencies; Lozinski, Hogbom and John for studies directed towards processes operating in periglacial environments; Bagnold for aeolian action; Grund, Mortel and Cvijic for karst studies; Zeuner, Pecsi and Budel for terraces in association with periglacial cycles and MacClintos, Martin, Fisk, Bryan, Ray, Hack, Thronthwaite, Smith and Condra for climatic geomorphology (Kumar, 1979).

In India, the geologists have to be given credit for initiating research in geomorphology. The work of Arogyaswamy (1967), Auden (1954), Chatterjee (1968), Dunn (1929), Krishnan (1953), Radhakrishna (1952), and Wadia (1937) are noteworthy. Geographers like Bagchi (1960), Chatterjee (1945), Chibber (1953) and Singh (1958, 1969, 1972, 1981) have pioneered work in geomorphology. On account of the adequecy of records and advent of different methods of study a lot of research works by eminent geologists and geographers are in progress. The range of their geomorphological works is now becoming much wider (Chatterjee, 1968, and Vaidyanathan, 1977).

Regional geomorphological studies of different parts of India at the macro, meso and micro levels have been made by geomorphologists. The contributions of Betal (1973), Pathak (1967), Kharkwal (1968), Kumar (1969), Singh (1969), Agarwal (1972), Arunachalam (1972), Singh (1975), Mukhopadhyay (1988) etc. may be mentioned. Many of the scholars have worked on fluvial geomorphology also. Basu (1976) studied some fundamental problems of meander formation with special reference to the Bhagirathi river. Kumar (1971) made significant geomorphic interpretation of the drainage and relief of eastern Ranchi plateau. He also studied the pairwise of basin area and stream length illustration relationship from the upper Burha basin. Kumar and Pandey (1977) made a numerical analysis of small drainage basins of Hazaribagh plateau region portraying their chief characteristics. Singh (1969) discussed the drainage pattern in the highlands of Chota Nagpur and traced their evolution and denudation chronology. Mache and Peshwa (1978) made an interesting photogeological interpretation of the controls on drainage in Gondwana and Bijawars of the Son valley, Shahdol district, Madhya Pradesh. A few works such as those of Kharkwal (1970), Satpathi (1972), Betal (1973), Sharma and Padmaja (1977), Kumar and Singh (1978) show effective use of quantitative methods in geomophic analysis.

Studies on floods, meanders, gullis and channel characteristic and slope analysis by Radhakrishna (1964), Niyogi (1968), Mukhopadhyay (1973), Chakraborty and Ghosh (1974), Das (1975), Pal and Bagchi (1975), Basu (1976), Kumar (1976), Dikshit (1976), Mukherjee (1983), and Pal (1983) are important and deserve mention.

Many scholars have also applied the empirical laws derived by Horton (1945), Strahler (1952), Schumm (1956) and Chorley (1969) in the study of drainage basin morphometry. The parameters have mostly been collected from topographical sheets by measuring lengths, breadths, areas etc. Satpathi (1976) has made similar study of the streams of Topa and Sulpi basins of Singbhum., while Singh (1986) has studied the morphological characterstics of some selected river basins of the Middle Narmada Valley.

1.5 IMPORTANCE OF THE STUDY

Manipur has remained a "*terra incognita*" so far as geological studies are concerned. The only information available so far are from the accounts of Oldham (1883), Pascoe (1912), Evans (1964) and from the reports of the piecemeal exploratory works of the Geological Survey of India. Most of the works confine to geological explorations and search for minerals. However, the GSI has started studies on geomorphological aspects within the Imphal valley and Loktak lake during the past few years. The main thrusts of the geographical studies so far carried out by various experts confine to political, demographic and economic aspects. While some of the physiographic description of the state have been made by regional geographers like Ansari (1976) and Singh (1982). Mukhopadhyay (1988) has portrayed a broad spectrum of geomorphological features of the landscape of Manipur. But these investigations can, by no means, lead to the micro level geomorphic descriptions of the landforms. Fluvial geomorphology is of paramount importance in the state of Manipur because of the fact that its surface is heavily drained by a number of intermontane river systems with potentials to bring about rapid changes in the landscape. Exposure of the land to torrential monsoon rain and heat of the sub - tropical to tropical climate on the one hand and the rapid rate of denudation of the watersheds on the other accentuate the rate of landform changes in the state. And such landform changes have tremendous implications in general ecology and environment of the hills and the central valley, the microclimate of the region and above all in the sociopolitical situations of the state. Therefore, geomorphological investigatios of different river basins of the state assume importance and claim priority.

It became imperative to consider a myriad of factors before the selection of the present area of study. Nevertheless not many factors can be singled out as influencing the author's choice. The following were considered while selecting the Iril river basin for the present work :

1. The length of the river and the area of its catchment basin are ideal for a work of this nature.
2. Its tributaries and watersheds are accessible from many sides which facilitated adequate ground survey work.
3. The forest cover, land use pattern and human habitations in its catchment and along its course have introduced features representative of various disturbed and undisturbed geomorphic forms of a typical river basin.
4. The river forms the largest tributary to the Manipur river (Chindwin - Irrawaddy system), holding a fairly significant position in the drainage system of the Manipur valley and for that matter, in irrigation and water supply potentials etc.
5. Since no such work on any of the river basins of Manipur has been taken up earlier, the present work is a modest attempt at terrain evaluation, watershed management and landuse pattern.

1.6 DESIGN OF THE STUDY

The study is divided into seven chapters. Chapter One, including this section, highlights the conceptual frame of the fluvial geomorphology and introduces the study area - the Iril river basin and its surroundings, besides presenting objectives, methodology, review of literature and importance of the study. Chapter Two deals with the general geology and structure of the study area, their impact on the landscape under the influence of the past and present day climate. This chapter also describes the soils, flora and fauna found in the area.

An assessment of the drainage characteristics has been presented in Chapter Three, which examines the drainage divides and their evolution, stream order and drainage pattern, stream frequency and confluence points, drainage texture and drainage density of the Iril river basin.

Chapter Four evaluates various morphometric characters of the Iril basin by studying its main tributaries such as Ithoi, Ihang, Ichai, Ekou, Leimakhong and Wari Lok etc. Order, number and length of streams, drainage texture and density, gradient, overland flow and shape of the basin have been studied in this chapter as they are intimately interrelated with each other.

Chapter Five analyses the slope and morphology of the landscape. Various methods of relief and slope analysis have been employed to verify the stability in relation to the ground level. Terrain evaluation, presented in this chapter, through different quantitative and cartographic techniques, portrays the areas of various relief region and process of basin development.

Chapter Six presents the geomorphic regions of the Iril river basin and analyses the land use pattern of the study area. Watershed management strategy for the basin has also been formulated in this part of the study.

Chapter Seven summarises the observations and findings in the concluding part of the study and suggests measures for watershed management and landuse improvements in the basin.

2

GEOLOGICAL SETTING

The geology of the eastern hills and the central valley of Manipur affect to much extent the form and features of the present landform of the Iril river basin. Geologically, as well as geomorphologically, Manipur is manifested with many interesting effects of structure, including tectonics and its history of evolution. "Manipur area may be regarded as a geomorphic model in Eastern India where the effects of interplay between structural, seismo - neotectonics and the changing fluvial, fluvio - marine etc. processess are well manifested" (Mukhopadhyay, 1988). Therefore, the study of its geology, tectonic structure, climatic conditions (past and present), flora and fauna are very important.

Manipur is divided into three distinct regions each with well defined character of its own (Singh, 1982), These divisions are :

a. The Manipur or Imphal valley.

b. The Manipur hills.

c. The Barak plain.

The Manipur or Imphal valley is underlain by an anticlinorium with its plunge towards the south (Oldham, 1883). The presence of rocks of marine origin of early and mid - Tertiary era suggests that the Manipur valley was formed either in the late Tertiary or in the early Pleistocene period due to tectonic movements associated with the Himalayan orogeny.

The Manipur hills, which enclose the central valley, are formed of Tertiary rocks and have ridges and valleys. The rivers of the eastern and southern parts of the hills drain into the Chindwin, a tributary of the Irrawaddy, and those of the northern and western parts flow into the Barak river which empties itself into the Ganga - Brahmaputra system in the Bay of Bengal.

2.1 GENERAL GEOLOGY

The geology of Manipur was first studied by Theobold (1873) and Oldham (1883). They found similarity between the sequence of rocks exposed in Manipur with the axials of the Arakan Yoma of Myanmar (Burma). Pascoe (1912)and Evans (1932) found that the rocks of the eastern hill ranges of Imphal are similar in lithological character with those of mallets - original Disang groups of rocks and to the west of Imphal with the Barails of Evans. Dayal and Duara (1963) also attempted a classification of the rocks of Manipur (Table 2.1).

Table 2.1

Rocks found in Manipur

Period		Rock formation
Recent to sub - recent	-	Alluvium
Oligocene	-	Barails
Cretaceous to early Eocene	-	Intrusive rocks Serpentinites
Cretaceous	-	Axials

The objective of portraying an account of general geology in connection with the study of geomorphological problem is to present the spatial distribution of the various rock groups and to indicate the role of these rock formations in the development of the present landform of the Iril basin. However, the geological information about Manipur is still incomplete. The Geological Survey of India (GSI) started mineral exploration in 1958, and geological mapping in some parts of the state in a piece meal manner during different field seasons. The geological maps procured from the different reports of the GSI for the present study area, the Iril basin, are shown in Fig. 2.1. The geological reports classify the rocks into argillaceous and arenaceous types. Based on the reports of several studies by Khan *et al* (1985, 1986), Sabale (1988), Shukla and Mohapatra (1988), Khan and Rao (1989), Shukla *et al* (1989), and on the finding of the present investigation, the stratigraphic succession with rock types has been worked out and presented in Table 2.2.

Khan and Jayaraman (1985) classified the Disang formation into upper Disang (wild flysch) and lower Disang based on the occurrence of olistoliths and olistostromal bodies of limestone and the tectonic slices of ultramafic. According to them, the upper Disang is with a chaotic assemblage of these olistostromal rocks of varying geological ages. Khan and Rao (1989) worked out various rock formations on the basis of the variation in lithological assemblage, fossil content and the mode of deposition. They subdivided the Barail group into Jenam formation, Laisong formation, Khoripok member and Melanze zone (sheared splintary shale with lenses of sandstone and lenticular bodies of limestone, brecciated chert and volcanics) and further Disang formation into Sanaching member and Thoubal member.

Table 2.2
Stratigraphic Succession and Rock Types

Rock Types	Period
Alluvium pebbles, gravels, soil etc. Thick bedded medium to coarse sandstone with less shale beds sometimes interbedded sequence of hard, compact and grey splintary shale, thin streaks of coal being common.	Recent Barail (upper Eocene to Oligocene)
Contact is gradational dominantly shale, siltstone, thinly bedded sandstone containing exotic blocks of ultramafic and limestone, Base not exposed.	Disang (upper Cretaceous to Eocene)

Figure 2.1 shows the rock groups within the Iril basin presenting a greater coverage by the Disang formation, dominantly of shale and siltstone. Sandstone beds are relatively less common and siltstone is generally soft and friable.

The shale is generally dark grey, yellowish in colour and splintary in nature. It is undergoing spheroidal weathering. The bedding planes are mostly obliterated due to incomplete stratification. Some fragmentary fossil remains of plants and leaves occur in the rocks.

The Barail group of rocks are conformably underlain by Disang

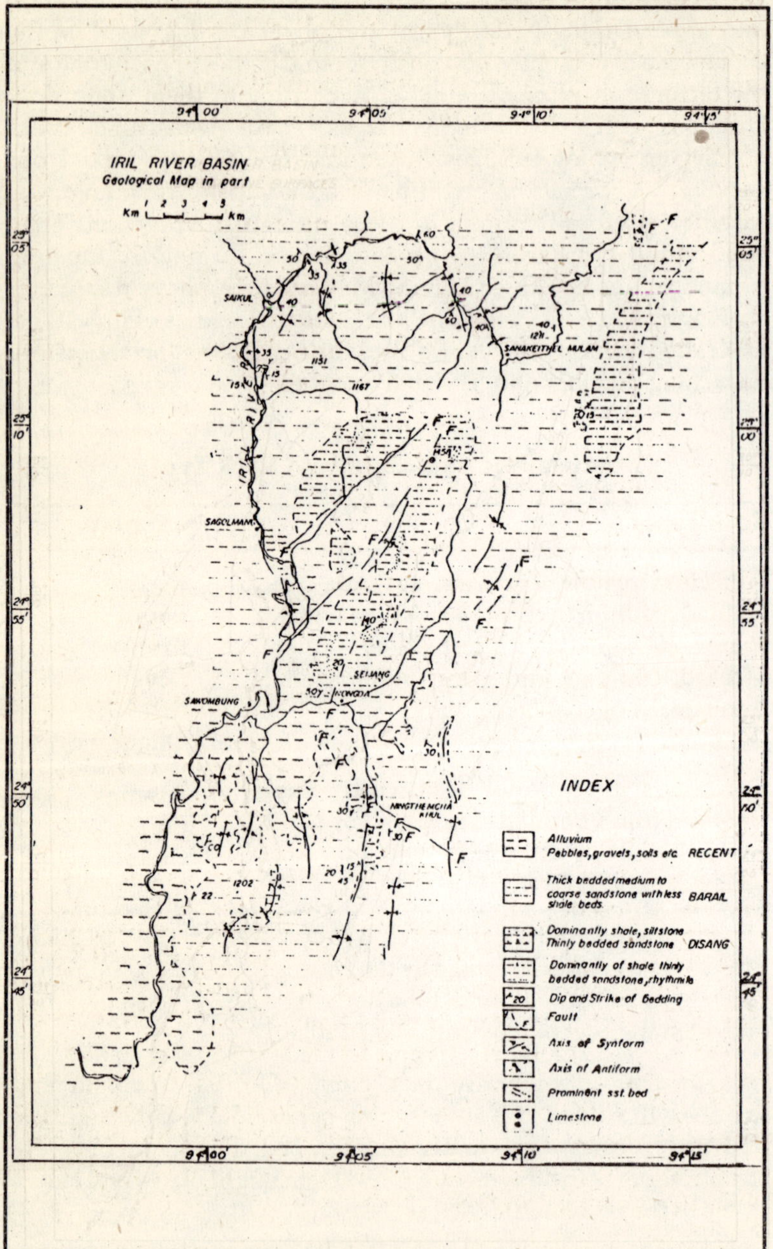
IRIL RIVER BASIN
Geological Map in part
Km 1 2 3 4 5 Km
SAIKUL
SANAKEITHEL MULAM
SAGOLMANG
SAWOMBUNG
SEIJANG
NINGTHEMCHA KHUL
INDEX
Alluvium
Pebbles, gravels, soils etc. RECENT
Thick bedded medium to coarse sandstone with less shale beds
BARAIL
Dominantly shale, siltstone
Thinly bedded sandstone
DISANG
Dominantly of shale thinly bedded sandstone, rhythmite
Dip and Strike of Bedding
Fault
Axis of Synform
Axis of Antiform
Prominent sst. bed
Limestone

Fig. 2.1

formation and both are considered to be representing geosynclinal facies (Khan *et al.* 1985). This formation is predominantly arenaceous in character with thick beds of sandstone, shale and siltstone. These occur at the core of the major synclines and the contact is mostly gradational with marks of gritty sandstone in the contact area. The shale and siltstone beds are similar to those of the Disang formation.

2.1.1 FOSSIL ASSSEMBLAGES

The main fossil assemblages of Disang and Barail are :

Gastropods : *Solariella* sp., *Ficus* sp., *Littorina* sp., *Turritella* sp., *Eclinochilus* sp., *Globularia* sp.

Pelecypods : *Glussus* (*Aralocardia*) sp., *Pilar* sp., *Cuspidaria* Sp., *Potamida* sp., *Gaestrocorbula* sp., *Nuculana* sp., *Arctica* sp.

Corals : *Stephanophyllia* sp., *Cycloliths* sp., *Tethocyathus* sp.

Foraminifers: *Nummulites* sp., *Globotruncana arena, Gumbelina* sp., *Striata* sp.

2.1.2 QUATERNARY SEDIMENTS

The Quaternary sediments are exposed on the differentially eroded basement of the Tertiary rocks and along the river courses. These sediments are of colluvial, alluvial and lumnic origin. Clay is the dominent constituent of the sub - surface sediment filled in the valley.

2.2 STRUCTURE

The hills of Manipur are part of the Indo - Myanmar mountain ranges that are peri - Indian continuation of the Alpine Himalayan chain (Mukhopadhyay, 1988). Structurally these hills are in the form of a series of open folded and doubly plunging synclines and anticlines. At many places, they are truncated by oblique high angle faults, bringing rocks of two diverse groups in juxtaposition. Thus, the older rocks, *i.e.* the Disangs, are exposed along the anticlinal cores, which are presently occupied by major river courses. The younger rocks of the Barail group are exposed along the ridge forming synclinal cores.

2.2.1 STRUCTURAL ZONES

Based on structural complexity, Manipur is divided into three structural zones by Basu and Ranga Rao (1964) These divisions are :

a. the central Manipur anticlinorium.

b. the moderately tightly folded zones of eastern Manipur hills.

c. the zone of close folding and thrusting in eastern Manipur.

The study area, the Iril river basin, occupies both the central Manipur anticlinorium and the zone of close folding and thrusting in eastern Manipur. Therefore, the characteristic features of both are encountered in the basin area.

Many of the primary structures such as bedding, current bedding of sandstone and siltstone, cross bedding and ripple marks, are indicative of the presence of shallow marine environment in a slowly subsiding basin. These are well exposed in this area. Biogenic structures are yet another primary structures, mainly present in the Barail sandstone in tubular, burrowed and flattened forms. These are also characteristic features of shallow marine environment.

This area has a complex history of deformation which is indicated by multiple folding, faulting and shearing. Srivastava *et al* (1976) have identified three phases of folding movements along the ultramafic belt of Manipur and Nagaland. These are more or less similar to the movements experienced by the Iril basin. They opine that the first phase of movements came with the early Alpine orogenic movements of the Eocene times, the second phase came with the late Alpine orogenic movements (Oligocene to Miocene) and the third phase coincided with the upliftment of the Mio - Pliocene sediments forming horst - graben like structures in the Indo - Myanmar ranges.

Three different phases of folding are inferred in the study area. The first generation of folding were in the from of localised tight isoclinal to open type of folds with NE-SW trending sub- vertical axial planes and low to moderate plunges (Khan and Rao, 1989). The second generation of folds were in the form of regional anticlines and synclines. These have affected the formation of the area and show various types of folding from open asymmetrical to tight symmetrical folds. A doubly plunging synform is found occupying the peak (1232 m). The hills such as Nungshigum, and Bungpiching, south of Monlam, are also synformal. The intervening valleys between most of the hills are antiformal, some of which are occupied by rivers Iril and Leimakhong. Almost all these folds are exposed and extended in more or less N-S to NNE - SSW direction (Khan and Rao, 1989).

The third generation of folds came in the form of broad warping trending almost E - W direction. The exposures of the third generation folds are seen in the prominent and persistent upstanding features of sandstone and siltstone in Mungching and the areas North, NE, NW and East of it. They display broad warping in the EW direction (Khan and Rao, 1989).

This basin area has a series of major strike faults which have restricted the domain of different phases of folding. The major faults of the area exert overall control on the topography and drainage of the basin area. They are discernible in the form of shear zones, slickensides, silicified and calcified rocks with brecciated structures and have subdued the changes in altitude of the strata and ommission of some beds or break in stratigraphic continuity. One of such major fault is the Iril fault. The Iril river flows along it in a NNW - SSE direction. It displays itself in the form of NE - SW trending Chanung fault between Keibi Khunou and Hudrom. This fault coincides more or less with the original anticlinal axis. The frequent alternation of dips, silicification and shearing along the fault is observed west of Leital Pokpi and Lamboikhul. Another major fault trends NE - SW and is 20km long. It displaces the southern part of the Iril fault. This fault has resulted in the obliteration of a part of the Barail formation from the Mungching, and brings the Barail and Disang formations in juxtaposition in the northern part. There is an abrupt change on the western side of the Mungching peak (1521 m).

Many other faults are also found in the study area. The Khamen Lok fault is 10 km long and trends NE - SW. It follows more or less the course of the major stream Khamen Lok and juxtaposes the Barail and Disang formations. The Nongda fault is 9 Km. long and trends NNE - SSW. It has displaced a portion of the fold closure to western limb (in the southern part of the Mungching ridge). It coincides more or less with a part of the doubly plunging synformal axis. The western block has been down thrown with an apparent displacement of about 1.5 km. A 14 km. long major fault follows the course of the Leimakhong. It trends NNE - SSW and coincides with the regional anticlinal axis. Effects of shearing and silicification is observed around Yasmongjong. Joints are found also in almost all the rocks of both the Barail and Disang series.

As the study area falls within the linear Assam - Arakan basin, it has very thick beds of Disang shales, Barail rocks and blocks of basic and ultrabasic rocks. The deposition of typical flysch sediments of the Disang appear to have been started during the late Cretaceous to the early Eocene period in a miogeosyncline. This is evident from the index micro fossils such as Globigerina, Globotruncana etc. present in the limestones. The presence of thick monotonous shale and thin siltstone beds amidst with few thick sandstone beds suggest an indication of gradual shallowing of the basin. Further, the presence of fossils in a restricted zone of Disang formation may be an indication of a short span of favourable climate for the growth of organisms. Some of the richly fossiliferous horizons in the Barail sediments, point towards the favourable conditions for the growth of various invertebrates and other organisms, and the deposition of the Barail sediments may have started in a shallow, gradually subsiding, estuarine basin. The presence of both large and minute forms of Discocydimids in a narrow zone (represented by grey carbonaceous sandy shale) during the deposition of the Barail group of rocks indicates existence of very shallow (5-10m) marine condition. Therefore, the temperature of water was more than that of surface water, *i.e.* it did not exceed 20^0 c. This condition was ideal for the growth of foraminifers in generally restricted upper parts of the photic zone with normal saline conditions. Environment for deep water deposition was prevalent during the upper Cretaceous period which is evident from the presence of limestone and chert.

2.3 CLIMATE

Ukhrul rainfall and temperature recording station is about 12 km east from the main catchment area of Iril basin, and on the south of the basin lies the Imphal meteorological station at a distance of about 10 km. Maram rainfall and temperature recording station is about 10 km north - west of the basin. Therefore, the climato - meteorological information used in the present study are based on the data procured from these three stations. However, before studying the present day climatic conditions of the basin, a brief study about palaeoclimate is also included to infer the intimate relationship between the past and present climate of the basin.

2.3.1 PALAEOCLIMATE

The erosional environments of the past in the valley of Manipur has been studied in geomorphological and geoarchaeological perspectives by many scholars which help in reconstructing the past climatic conditions of Manipur. From the marine environment under the Tethys sea, the land was raised slowly and exposed to the warm climate of the Tertiary period. It appears that the land escaped the Pleistocene glaciation of the early Quaternary period and remained cool and dry (Singh and Shyamananda, 1988). Palynological studies of the surface and sub - surface sediments of Loktak lake area (Ray and Chanda, 1989) also showed a cool and humid condition as indicated by the lowermost sediments at a depth of ±2.40 metre supporting the above climatic condition. Further, there is no geological evidence of Quaternary glacial or periglacial climate upto the elevation of about 1500 m above the mean sea level.

In fact, ferrugineous (oxidised) cement and matrix of gravel formation in the region suggest a sub - tropical Quaternary (Joshi *et al*, 1987). Therefore, the climate in the late Quaternary was warm and humid which is also supported by the study of pollen in the next higher strata at a depth of ±1.20 metre (Ray and Chanda, 1989). This indicates a tropical to sub tropicl forest condition and the palynological study of the upper surface showed tropical rain forest which has not changed drastically since then.

2.3.2 PRESENT CLIMATE

The basin area enjoys a humid tropical to subtropical climate. The general temperature is modified by the relief of the hill ranges. Again rainfall is one of the most important climatic parameters in shaping the landforms of a region. Rains of high intensity cause acceleration in the process of denudation in the deforested parts of the basin. The amount of rainfall increases from the valley towards the hill region. The tropical to sub - tropical climate largely prevail in the relief region upto 1,800 m, and the higher regions enjoy more or less a temperate climate.

Annual rainfall is sometime as low as 837 mm and sometime as high as 2047 mm. The temperature ranges between 1.5°C and 33°C. Average relative humidity is between 17 per cent and 100 per cent. At the level of Imphal valley, the rate of evaporation ranges between 1.8 mm in January and 4 mm in May. Duration of bright sunshine has a variation of 3 hrs./ day in July and 9 hrs./ day during January and the surface level pressure ranges between 900 mb and 930 mb in the valley portion.

Table 2.3
Climato-meteorological data of three stations around Iril basin

a. Relative humidity (p.c.)

Station	Jan	Feb	Mar	Apr	May	Jun	Jul	Aug	Sep	Oct	Nov	Dec
Maram	65	62	61	68	75	84	80	82	80	82	75	78
Imphal	70	68	66	82	90	88	90	90	85	85	82	80
Ukhrul	82	80	76	84	82	90	92	96	100	88	82	80

b. Rainfall (mm)

Station	Jan	Feb	Mar	Apr	May	Jun	Jul	Aug	Sep	Oct	Nov	Dec
Maram	05	10	80	50	160	130	200	190	185	180	08	06
Imphal	04	08	80	100	180	200	260	180	110	245	06	05
Ukhrul	04	20	85	60	90	330	405	275	100	320	08	10

c. Max. temperature (°C)

Station	Jan	Feb	Mar	Apr	May	Jun	Jul	Aug	Sep	Oct	Nov	Dec
Maram	20	21	24	28	32	31	32	22	30	29	20	18
Imphal	21	23	26	32	31	31	32	33	31	32	25	20
Ukhrul	20	19	25	28	28	30.5	33	30	27.5	27	24	24.5

d. Min. temperature (°C)

Station	Jan	Feb	Mar	Apr	May	Jun	Jul	Aug	Sep	Oct	Nov	Dec
Maram	2	3	6	10	10.5	10.5	11	7	10.5	7	8	4.5
Imphal	0	4	7.5	10.5	16	20.5	22	16	11.5	8.5	4	0.5
Ukhrul	1.5	3	2.5	8	10.5	12	14	12.5	12	9.5	8	8.5

Figure 2.2 shows the variation of average value of temperature (maximum and minimum), humidity and rainfall at Ukhrul, Imphal and Maram which are in close proximity to the Iril basin

area. Table 2.3 shows the climatic data recorded at these stations for the period of ten years (1979-90).

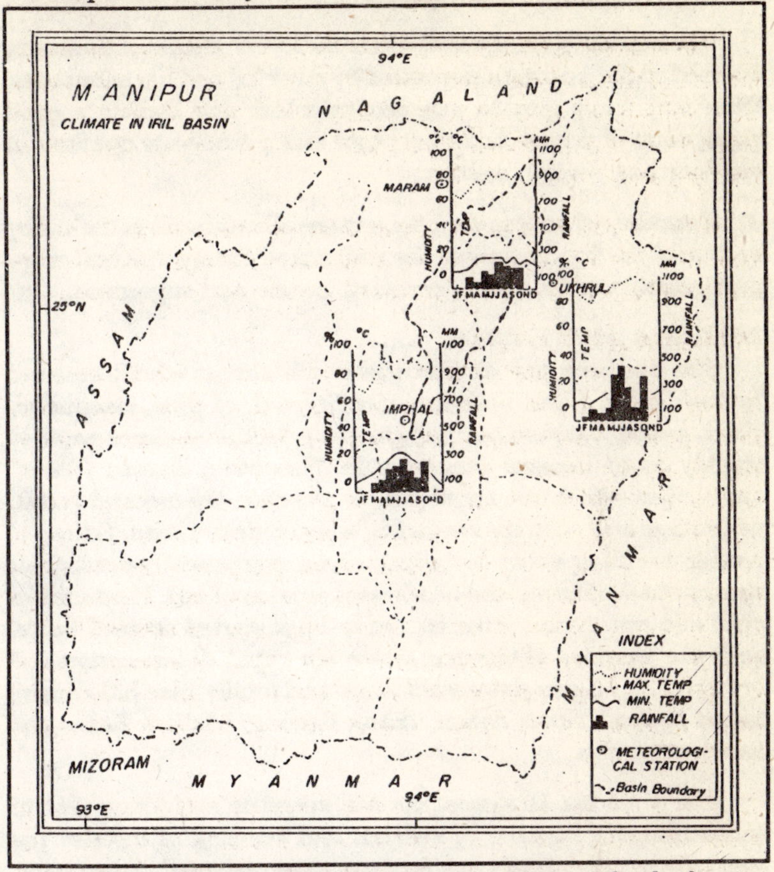

Fig. 2.2 Map of Manipur showing Meteorological environment of Three stations

2.4 SOILS

The soils of the Iril river basin belong to two groups - residual soil in the hilly section and the transported soils in the valley portion. The ferrugineous red gravelly soils, which belong to the non - laterized red soil (*Ultisol*) group of the residual soils contain small gravels interspersed with sand and silt. These soils are generally shallow in depth and yellow and red in colour. They have medium to strong acidity, a fair amount of nitrogen and are fairly rich in phosphate and potash. On the north - western margin of the basin these soils contain less gravels and become loamy in texture and granular in structure.

The alluvial soils (*Entisol*) cover the lower section of the basin. Formed of the materials deposited by river Iril and her tributaries, these soils have grey to pale brown colour and contain a good proportion of potash and phosphate and a moderate quantity of nitrogen and organic matter.

Both the red soils in the hills and the alluvial soils in the valley section of the Iril catchment are well suited for agricultural crops particularly paddy, maize, mustard, potato and sugarcane.

2.5 FLORA AND FAUNA

The northern part of the basin is still almost wholly covered by lush green forest with luxuriant growth of pine, cinammon, plum, peach, bamboo and tall grasses which sometimes grow so thickly as to became inaccessible. However, due to "*Jhum*" cultivation which the tribal people practise, the natural forest, vegetation and wild life are facing wanton destruction. Felling of timber yielding trees for commercial purposes causes great damage to the forests and accelerates erosion of soil. Plantation of pine and eucalyptus trees etc. have been started around Saikul with the purpose of slowing down the rapid deforestation and soil erosion. Highly deforested areas and totally bare hillocks are found around Ekou, Saikul, Ihang Karong, Malkon Bazar and Sanakeithel area.

The wild life is almost on the verge of extinction due to indescriminate hunting of animals and burning of forests. The common wild animals are barking deer, leopard, bear, tiger, jackals, wild boar etc. Among the animals, the barking deer (*Saji*) is the commonest object of hunting by the inhabitants of the hilly regions.

3

DRAINAGE CHARACTERISTICS

The denudational work and its transformation in different stages of its evolution reveal a systematic account of the origin and stages of erosion at surfaces of the Iril river basin.The basin has developed under a humid climatic condition receiving excessive monsoon rainfall, which enhances the erosion and denudation of surface landforms in the area. The Iril river and its tributaries are at present incised in small valleys and some parts of its channel bed show complications with alluvial and gravel terraces (Plate 3.1) thus bearing the evidences of Tertiary elevation and denudation and Quaternary cycles. With many interruptions, the river has passed through several cycles of extension with interruption and has the present maturity stage.

The Iril river is the master (main) stream of the Iril drainage basin. It has a number of main tributaries such as the Ithoi Lok, Ihang Lok, Ekou Lok, Leimakhong, Ichai Lok etc. The Iril, its main tributaries and sub - tributaries, the rivulets and rills constitute the Iril river system. The part of the land surface, the rain water of which is drained out by the Iril river system, may be called the Iril catchment or drainage basin (Fig.3.1). This drainage basin has an area of 1,285 sq.km.

The main drainage axis of the Iril river runs in NE - SW directions. The Iril river, with its right and left bank tributaries, flows first in the hilly terrain up to village Sagolmang, where it debouches onto the Manipur valley. On emerging over the plain, the river looses much of its energy and its whole character is changed. It begins to flow in a zigzag course forming many meanders. The Iril discharges its water into the Imphal river at Lilong. River Imphal, further south, collects water of Thoubal, Sekmai, Khordak and Khuga rivers and emerges as Manipur or Achouba river. This master stream after negotiating a gorge and

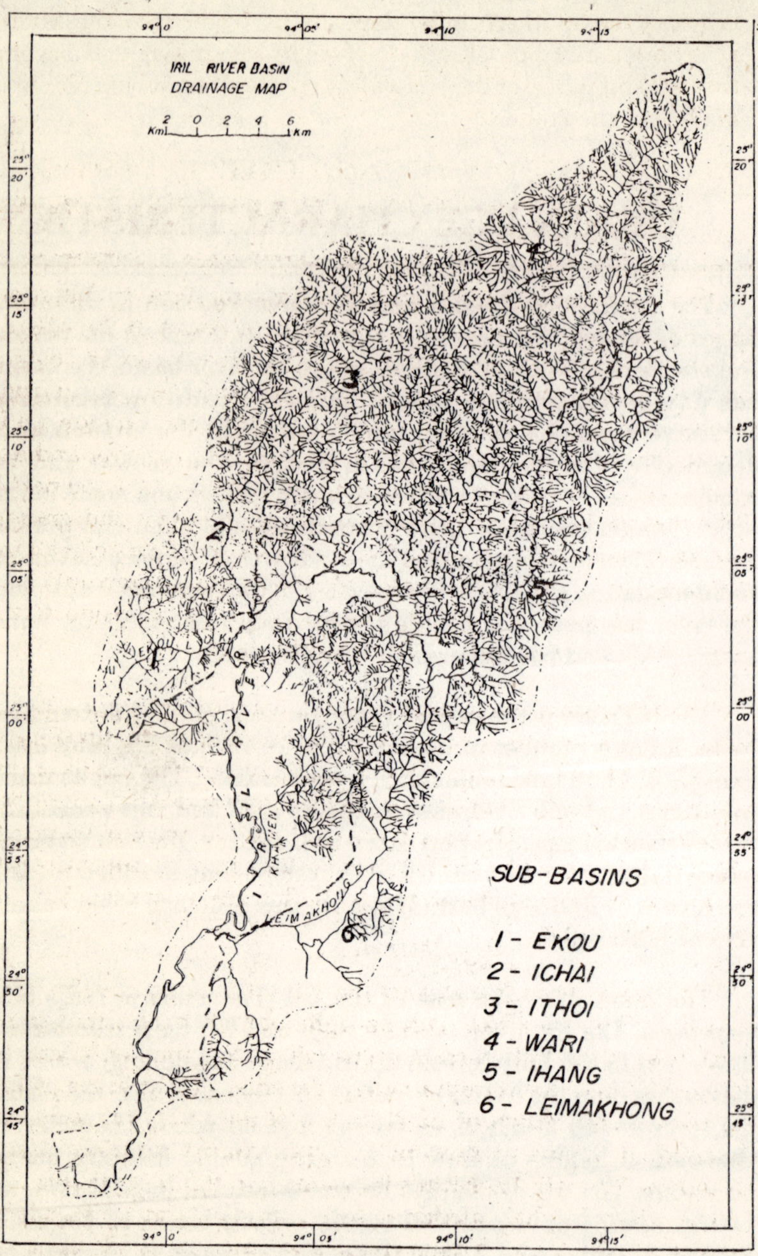
IRIL RIVER BASIN
DRAINAGE MAP
2 0 2 4 6
Km Km
SUB-BASINS
1 - EKOU
2 - ICHAI
3 - ITHOI
4 - WARI
5 - IHANG
6 - LEIMAKHONG

Fig. 3.1

crossing the southern hills, deposits its water into the Myittha river, a tributary of Chindwin river in Myammar and becomes the part and parcel of the Irrawaddy river system which ultimately pools into the Andaman sea.

3.1 DRAINAGE DIVIDES AND THEIR EVOLUTION

Drainage divides, water divides or watersheds act as water parting between adjacent rivers or drainage basins of equal or unequal sizes. They, thus, act as outer boundaries of major or minor river basins. Nothing can be more interesting in the study of subaerial denudation, which shapes the drainage basins, than the identification, subdivision and re - arrangements of the divides (Davis, 1954). The water divides are of two categories depending upon the nature of their surface of configuration. These are (1) primary divides with undulating surface configuration, and (2) secondary divides with much less undulations. The former are generally found in peneplained surfaces. As the geomorphic cycle, better known as geographic cycle, passes from one stage to another, the divides migrate towards the weaker streams because stronger streams degrade and widen their valleys rapidly, whereby a number of divides and weaker streams are wiped out. Thus, the divides, and with them the streams of the river basin, go on decreasing in number. The layout of the divides determines the pattern of the drainage channels. Therefore, the divides in no way are less important than the drainage channels. They are the surface manifestations of the landforms at any point of time in the history of evolution of landscape. Therefore, they are the best aid to the appreciation of landscape.

Within the Iril river basin, the primary divides separate the basins of the Ihang, Ithoi, Ekou, Khamen, Leimakhong etc. The secondary divides demarcate the basins of their tributaries. Thus, there are basins within basin, and their number depends upon the number of streams and rivulets occurring in the Iril river basin. The basin has directed the divides (mostly ridges) in its hilly reaches. The undulation gradually decreases towards the south, and the interfluves or interstream areas are almost uniform especially in the Manipur valley.

The history of evolution of a drainage basin and its morphology are closely related to geology, *i.e.* the type of rocks and their structure, in its broad connotation, the various sub -

aerial processes active in the region and the stage of evolution reached. In a region of hard and resistant rocks, the rivers generally select the lines of weakness in the basement rocks for carving out their valleys. Most often the rivers do not disregard the structure on which they flow. On the other hand, in a region of homogeneous rock, they flow in any direction in sympathy with the regional gradient. There is absence of structural controls in such areas.

Depending upon the susceptibility to erosion and structure of the rocks on which the drainage basin is formed, the amount of rainfall, the rapidity with which the fluvial processess operate and the stage of evolution reached, the rivers undergo branching. The basin has master stream and many major and minor tributaries, which in turn have smaller tributaries, rivulets and rills.The drainage map, thus, comprises of drainage channels of various sizes. With the progress of the cycle of erosion, drainage is integrated and the number of drainage channels goes on decreasing. Thus, the state of branching of streams may throw light on the nature of the rocks, their structure and the progress of the cycle of erosion.

Although the geology and landform structure have done much work to attain the drainage systems, climatic factors still exert a great influence upon the present erosional processes which, in turn, contribute much in determining the present drainage pattern of the Iril basin. However, the Iril river and the hill slope processes provide the important theme of the multicyclic evolution especially by its drainage pattern in the upper reaches of the river itself. Some of the distinct geomorphic units seen to correlate with that chalked out for the Iril basin, where regional uplifts in its past were accompanied by warping, faulting or tilting of the erosional surfaces. The upper middle portion of the Iril river, initiated over a structural area with a slope towards the south and south - west direction, has progressively down cut upto the valley portion. The differences in the relative relief and altitude between the present river valley and the summits of the surrounding highlands is due to the initial uplift which this basin has experienced.

The terrain of the basin has been controlled by the Iril river and its tributaries. It appears that the hills or the highlands, present in the basin, are the product of earth sculpture. In course of time, they have been reduced as the present day highlands, and most of the vast landmass brings to lower level during the periods of uplifts. Finally, it can be stated that there is a close relation between the patterns of structure and Iril drainage with its sub - tributaries in the basin itself. Some portions of the Iril drainage system developed preferentially along the line of faults and joints specially in week impermeable rocks.

3.2 STREAM ORDER

The streams of a drainage basin are classified as belonging to a number of stream orders on the basis of the size of the drainage channels and the generation to which they belong. By stream order is meant the portion of a stream in a hierarchy of tributaries (Leopold, *et. al.*, 1964). The smallest stream of the drainage basin, which has no feeder, is put in the category of first order stream. The union of any two first order streams generates a channel of second order stream, and with union of two second order streams comes into existence the channel of a third order stream and so on (Singh, 1960). The master stream is the last in the series of stream orders. Thus, the order number goes on increasing from the smallest to the biggest stream of the drainage basin. The stream order of a particular stream is its serial number, which begins from the smallest stream. Stream order thus gives a quantitative basis for comparison of the degree of development in the drainage nets of comparable size (Horton, 1945). It is an obvious conclusion that the numerical strength of the streams goes on decreasing with the increase of their hierarchical order.

3.3 DRAINAGE PATTERN

The plan or design that emerges on a drainage map on account of the assemblage of drainage channels is called drainage pattern. What type of drainage pattern will develop in a drainage basin depends, among others, on the types of rocks and their structure. For example, a dendritic pattern emerges in an area of homogeneous rocks, a radial pattern on a domed structure, and a trellis pattern on a folded structure. Thus, the types of rocks and their structures can be inferred with a reasonable amount of accuracy by perusal of the drainage pattern developed on a drainage map.

3.3.1 SIGNIFICANCE OF DRAINAGE PATTERN

The drainage pattern can reflect the influence of initial slope, resistivity to erosion of rocks, structural control, recent geologic and geomorphic history of the drainage basin. Because drainage patterns are influenced by so many factors, they are extremely helpful in the interpretation of geomorphic features and the study of them represents one of the more practical approaches to an understanding of structural and lithologic control of landform evolution (Thornbury, 1985). Therefore, the detailed study of the drainage pattern of the Iril basin was taken up to evaluate its geomorphic features and its structural and lithologic controls.

3.3.2 TYPES OF DRAINAGE PATTERN

For the purpose of the present study the drainage channels and contours were traced from the 1:50,000 topographical sheet Nos. 83K/3, 4, 7, 8; 83G/ 16; 83L/ 1 and 83H/13 and 14 (Fig. 3.1) for a detailed analysis of the morphology of the Iril drainage basin. The drainage pattern that emerged corresponds with dendritic or sub - dendritic pattern with irregular branching of streams, and rectangular or angulate pattern, while in some residual hills radial pattern of drainage are found.

a. Dendritic Pattern

The most common drainage pattern found in the area is the dendritic pattern. This pattern of drainage is suggestive of little adjustment to underlying structures. Generally dendritic pattern develops upon rocks of uniform resistence and found upon nearly horizontal sedimentary rocks or in area of massive igneous rocks but may be seen on folded or complexly metamorphosed rocks, particularly where imposed upon them by superposition (Thornbury, 1985). In such pattern of drainage, the tributary joins the trunk stream at acute angles, and it is regarded as the normal pattern of drainage. However, in this context, it may be visualized that the drainage of Iril basin is mainly of dendritic pattern. Undoubtedly, the diastrophism and erosion have influenced the basin in attaining the present form/ shape, which may be seen as a complex form of landform supported by various cycles of erosion. Within the broad dendritic pattern of streams, some of the sub - tributaries such as Kumpareng Lok, Selchi Vadung, Shorbung Lui, the upstream part of Tusom Lok and middle part of Leimakhong are all having more or less straight courses, which appear to be the effect of structural control, although dendritic pattern is generally associated with the areas of less structural

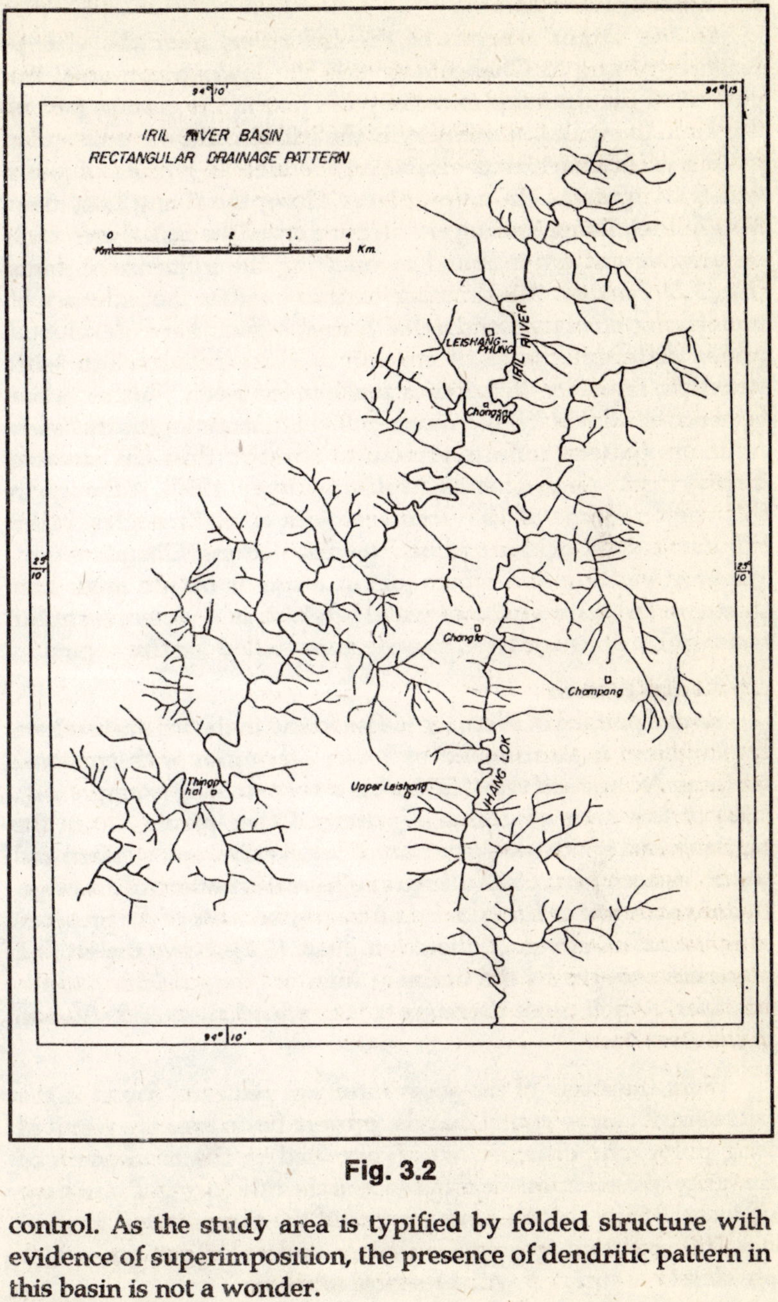

Fig. 3.2

control. As the study area is typified by folded structure with evidence of superimposition, the presence of dendritic pattern in this basin is not a wonder.

b. Rectangular Pattern

In the upper reaches of the Iril river, near the village Leishangphung to Changshang and its downstream area, the pattern of the drainage is rectangular just in the eastern part of this area. Ihang Lok, a tributary of the Iril, also shows rectangular pattern of drainage thus exhibiting the area as formed of joints and fault systems. In many places along the Ihang Lok, near Sanakeithel, Ihang Karong and upper part of the Saikul area such rectangular pattern is found manifesting the influence of joints (Fig. 3.2). Further, this drainage pattern exhibits the influence of underlying structure and relief forms, which have developed under different cycles of erosion in close relationship with structure, specially the joints or faults in the basin. Further, it has been accepted widely by a number of geomorphologists that such drainage pattern reflects structural control, thus the streams display right - angled bends (Mukhopadhyay, 1980). Although in this region, some of the drainages join at right angles, other tributaries such as Maring Lok, Haochu Vadung, Chandare Lok, Vaphrey Vadung and others join at acute or obtuse angle and therefore the term angulate pattern, which is a varient term for rectangular drainage type, may be more suited for this region.

c. Radial Pattern

Radial pattern of drainage is also found in the Iril basin where the highland is surrounded by lower elevations with erosional surfaces. Nongmaiching (1,583m.) is a very striking example (Fig. 3.2a) of this type of drainage pattern. Other places, *i.e.* in the northern and south - western part of Sanakeithel and western and south - eastern part of Saikul area also have this pattern of drainage. Radial pattern of drainage is found mainly because of the presence of erosional slopes in all direction. Plate (3.2) shows the eroded materials brought by the drainage lines of Nongmaiching to the Iril river, which cause damages to its embankment and siltation to the river bed.

From the study of the above drainage patterns, found in the Iril basin, it can be stated that the present basin area is a result of long polycyclic changes, which resulted in the attainment of maturity of dissection for difference of the lithology and structure of rocks. Many factors have produced the complicated form of dendritic pattern, rectangular pattern and radial pattern, which are closely associated with erosional surface.

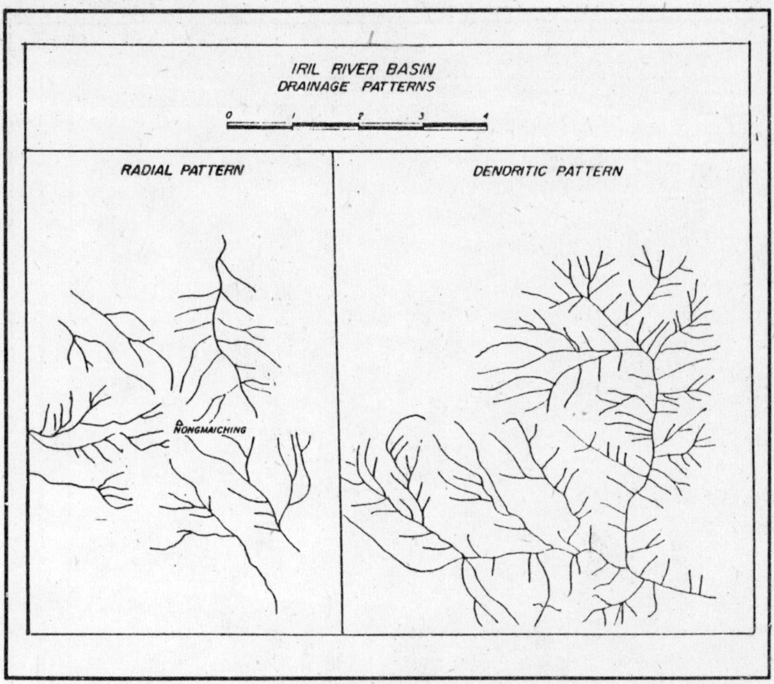

Fig. 3.2a

3.4 NUMBER OF STREAMS AND THEIR FREQUENCY

The total number of streams found in a drainage basin as a whole is important no doubt, but of greater importance is their number per unit area, and their frequency of occurrence in different parts of the drainage basin. This variation throws light on many important aspects. In order to obtain the number of streams per unit area, a network of vertical and horizontal grid lines were superimposed at one kilometre interval on the drainage map. The drainage map was thus divided into 1285 squares, each of one square kilometre in area. Taking the squares one by one, streams and parts of streams (stream segments) occuring in each of them were counted and the numbers indicated at their centres. As the squares are one square kilometre in area, the figures represent the number of streams per square kilometre, the stream frequency. Isopleth lines were then drawn using the frequency figures in the squares. After selecting suitable class interval and frequency ranges, the figures were entered in a frequency table and the number of squares in each of the frequency ranges were computed.

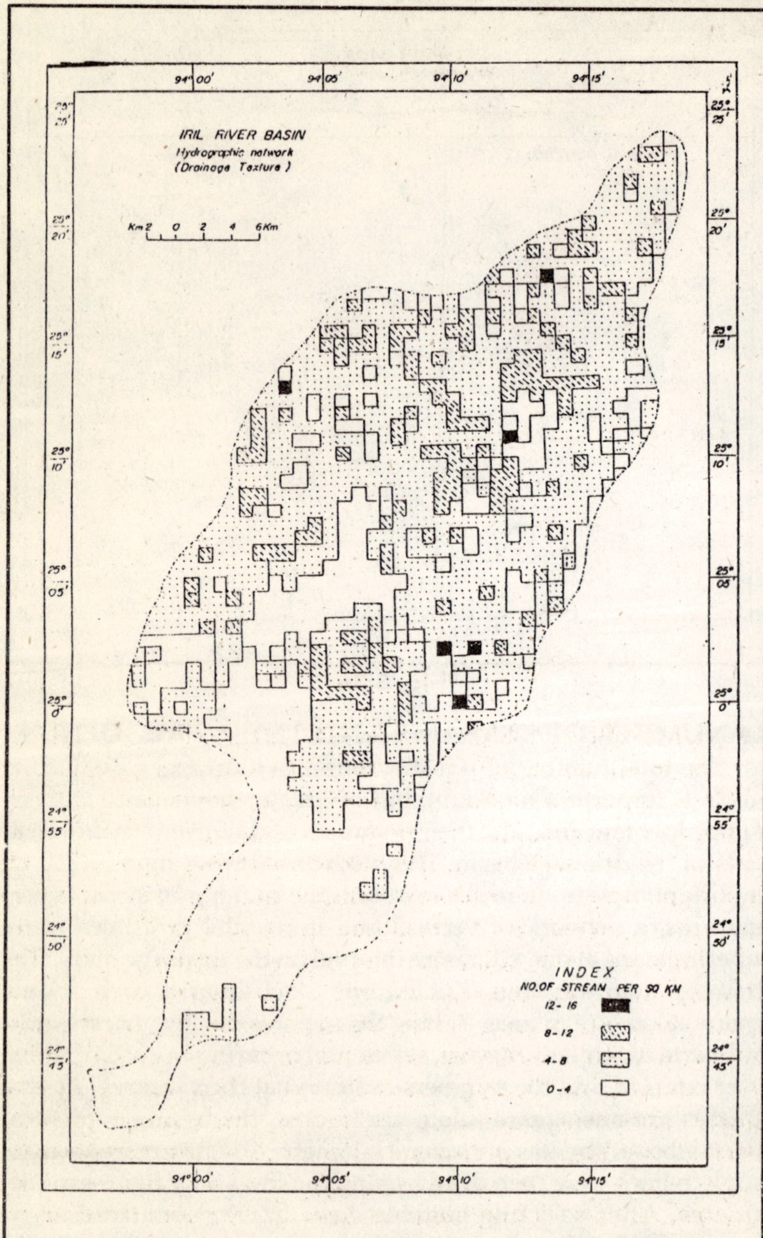
IRIL RIVER BASIN
Hydrographic network
(Drainage Texture)
Km 2 0 2 4 6 Km
INDEX
NO. OF STREAM PER SQ KM
12-16
8-12
4-8
0-4
94°00'
94°05'
94°10'
94°15'
25° 25'
25° 20'
25° 15'
25° 10'
25° 05'
25° 0'
24° 55'
24° 50'
24° 45'

Fig. 3.3

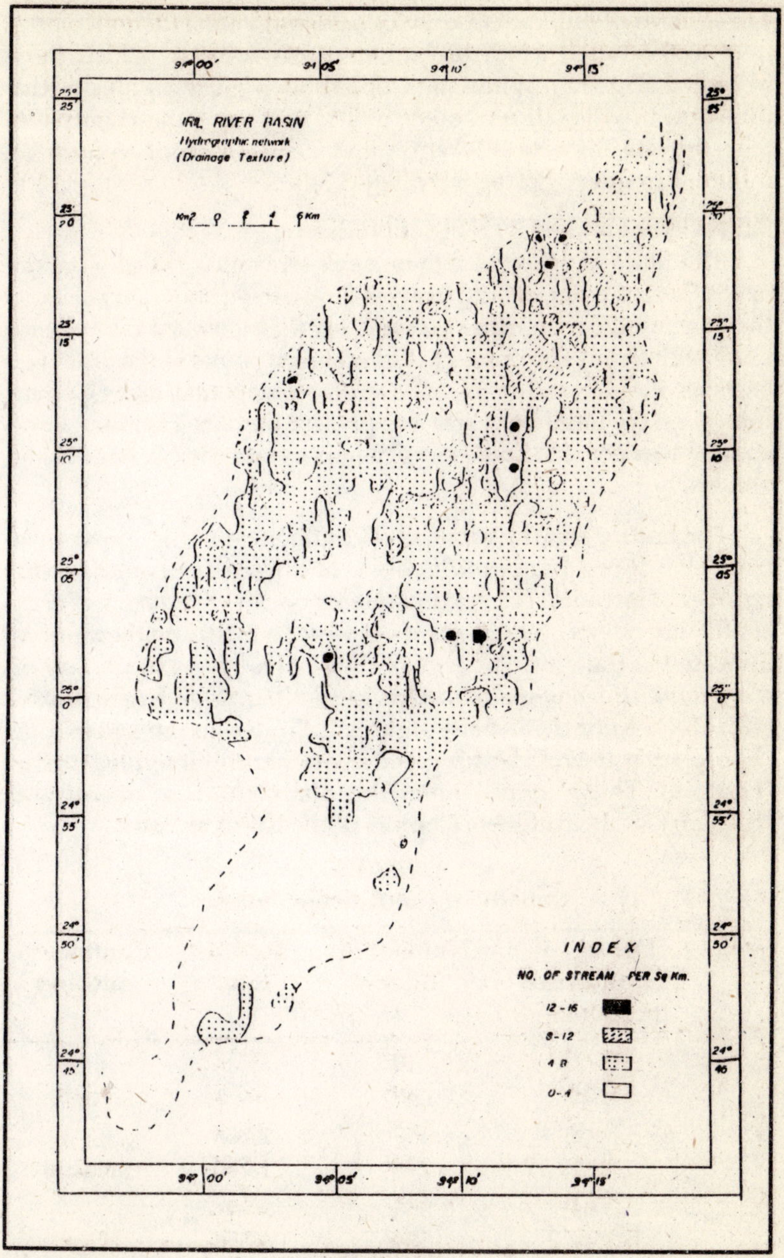
IRIL RIVER BASIN
Hydrographic network
(Drainage Texture)
INDEX
NO. OF STREAM PER Sq Km.
12 - 16
8 - 12
4 8
0 - 4

Fig. 3.3a

The squares of different frequency categories were identified on a map and shaded according to the categories to which they belonged (Fig. 3.3). While the isopleth map gives an idea of the direction in which the number of streams vary, the choropleth map depicts the actual location and distribution of squares of different frequency categories (Fig. 3.3a).

3.5 NUMBER OF CONFLUENCES

The place of union of two or more streams is called junction or confluence point. As the streams flow away from their sources, they encounter with other streams on their onward march and unite with them. The number of confluence points in the drainage basin as a whole and in its different parts depends, among other things, on the number of streams present therein. The number of confluences per unit area is a very important index in river basin studies.

For finding out the number of confluences in the squares we used the drainage map which we had used for counting the number of streams. Taking the squares one by one the number of confluence points occurring in them were counted and the number indicated at the centre of each of them. This gives the density of confluences per square kilometre. Isopleth lines were then drawn with the help of the density figures indicated in the squares. A choropleth map was also prepared in the manner described earlier (Fig. 3.4). These maps depict the distribution of density or frequency of the confluence points in the Iril river basin.

Table 3.1
Density of Confluence Points

Sl.No.	Density of Confluences per sq. km.	Number of Squares	% of the total	Confluence category
I	10 - 12	10	0.78	
	8 -10	48	3.74	High
II	6 - 8	155	12.06	
	4 - 6	232	18.05	Medium
III	2 - 4	306	23.81	
	Below 2	534	41.56	Low
		1285	100.00	

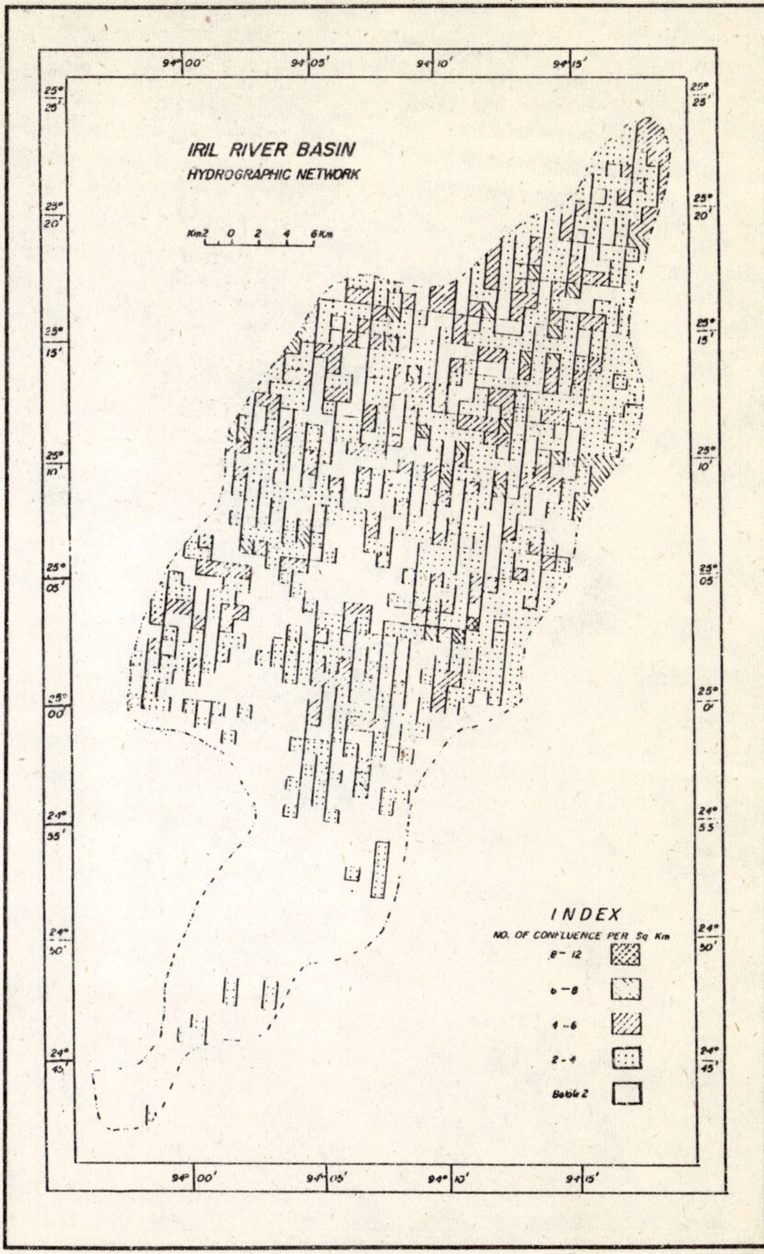
IRIL RIVER BASIN
HYDROGRAPHIC NETWORK
Km 2 0 2 4 6 Km
94° 00'
94° 05'
94° 10'
94° 15'
25° 25'
25° 20'
25° 15'
25° 10'
25° 05'
25° 00'
24° 55'
24° 50'
24° 45'
INDEX
NO. OF CONFLUENCE PER Sq Km
8 – 12
6 – 8
4 – 6
2 – 4
Belov 2

Fig. 3.4

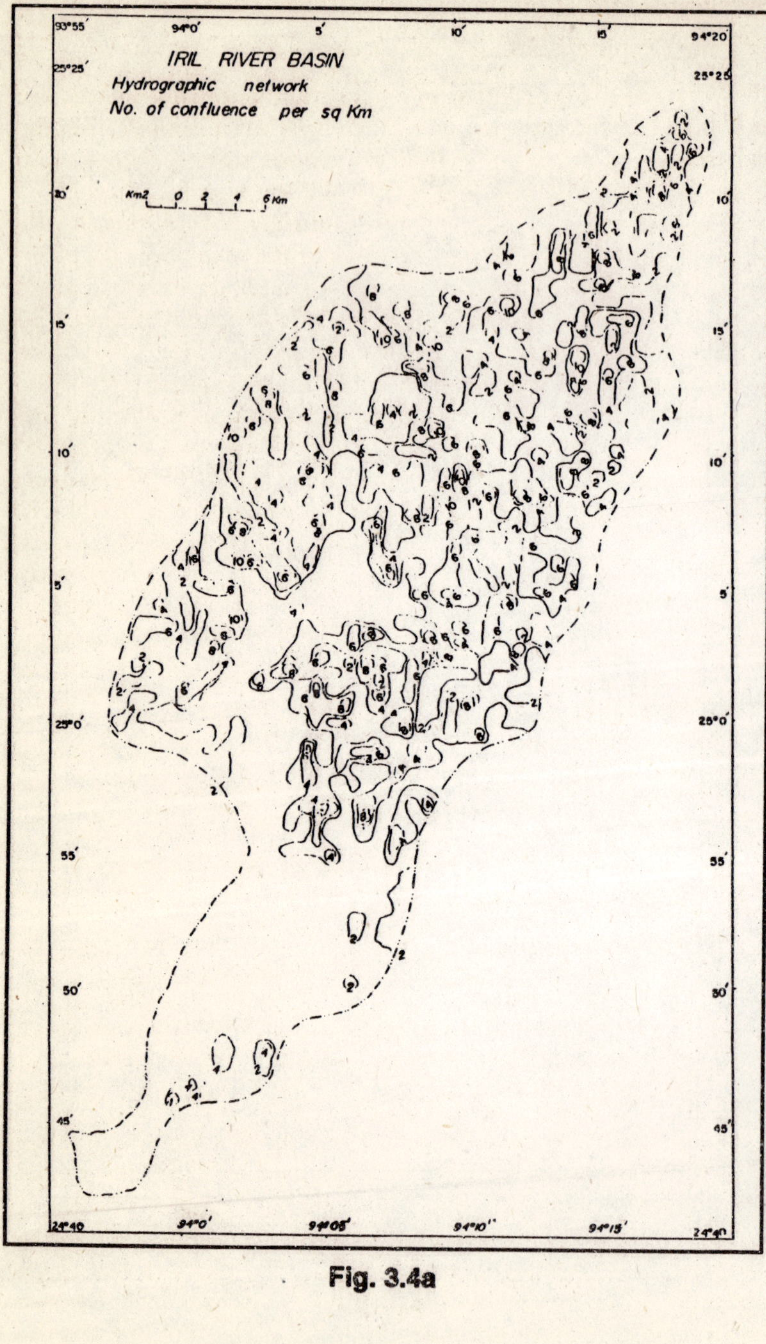
IRIL RIVER BASIN
Hydrographic network
No. of confluence per sq Km
Km2 0 2 4 6 Km
93°55′
94°0′
94°20′
25°25′
25°0′
24°40′
94°05′
94°10′
94°15′

Fig. 3.4a

The density of confluences, *i.e.* the number of confluences per sq. km. have been grouped into six categories in Table 3.1 For the convenience of description these six groups were combined into three *viz.,* high, medium and low. It will be seen from Table 3.1 and Figs. 3.4 and 3.4a that the largest number of confluences (10 - 12) are found in only 0.78 per cent of the total area of the Iril basin while the smallest number of the confluences are found in more than two - fifths (41.56 per cent) of the total area of the basin. The range is quite significant. There is predominance of low confluence density (upto 4 per sq km) in the whole of the central part of the Iril basin, where the Ihang Lok and Ithoi Lok meet with the Iril river and around the mouth of the latter. These two parts of the basin together claim slightly less than two - thirds (65.37 per cent) of the total area of the Iril basin. High confluence density (8 - 12 per sq km) is found in the border lands, *i.e.,* in the northwestern and south - western hilly parts of the basin around Laishan peak (2260 m.) and Mungching. These two parts of the basin together cover a very insignificant proportion (4.52 per cent) of its total area. The density of confluences varies between 4 and 8 per sq km in the remaining parts of the basin comprising less than a third (30.11 per cent) of its total area.

An examination of the maps showing distribution of number of streams and confluence points reveals that :

(1) the number of streams and confluences are larger in the northern and south-eastern part of the Iril basin,

(2) the narrow and elongated ridges and scarps have also larger number of streams and confluences and

(3) the central and southern parts of the Iril basin have comparatively smaller number of streams and confluences.

There appears to be a close relationship between the number of streams and confluences on the one hand and the intensity of erosion on the other. We presume so because greater amount of erosion is taking place and a greater dissection is observed in the northern, north-western and south-western parts of the basin especially in the narrow and elongated ridges and scarps,where the streams and confluences are found in larger number. On the other hand,the impact of erosion is minimal and the amount of dissection is the least in the central part of the Iril drainage basin around where the Ihang lok and Ithoi Lok meet with the Iril, and in the southern part around its junction with Imphal river where the streams and confluences are found in very small numbers. Thus, we may conclude that larger the number of streams and confluences, greater is the amount of dissection of the land surface.

3.6 DRAINAGE TEXTURE

Drainage texture is important to geomorphic analysis. By drainage texture is meant " the relative spacing of drainage lines" (Thornbury,1957). It refers to the distance separating the drainage channels. As a matter of fact, "it includes both drainage density and stream frequency" (Horton, 1945). Drainage texture is determined by rainfall, temperature, permeability and structure of the rocks on which the drainage basin develops (Satpathi, 1981). Hydrographic network or the drainage texture map of the Iril river has been shown by choropleth and isopleth (Figs. 3.3 and 3.3a). The isopleth map shows the direction in which the drainage texture varies (Fig.3.3a). After selecting a suitable class interval, four stream density ranges were formulated and the figures of the squares were distributed among them. Table 3.2 shows the number of squares falling in the four stream density ranges. For convenience of description, the second and third stream density categories were combined to reduce the categories to three - fine, medium and coarse. The choropleth map shows these three categories (Fig.3.3)

Table 3.2
Density of Streams

Sl. No.	Number of streams per sq km	Number of Squares	% of the total	Drainage texture
I	12 - 16	14	1.09	Fine
II	8 - 12	249	19.38	
	4 - 8	611	47.55	Medium
III	Below 4	411	31.98	Coarse
		1285	100.00	

It will be seen in Table 3.2 and Fig. 3.3 and 3.3a that in only 1.09 per cent of the total area of the Iril drainage basin the texture of drainage is fine, meaning thereby that the drainage channels are closely spaced, 12 to 16 of them occurring over a distance of one km. The interstream areas are found to be narrow. The parts of the basin, where this situation is obtained, are found scattered occurring mainly in the northern and south - eastern parts. The areas of medium drainage texture are overwhelming and pre - dominant, claiming over two - thirds (66.93 per cent) of the total area of the Iril basin. Some 4 to 12 streams occur over a distance of one km in this part of the basin. The drainage texture is coarse, with less than 4 streams over a distance of one km, mostly in the central and southern parts of the Iril basin. These two areas together comprise slightly less than a third (31.98 per cent) of the total area of the basin.

The following facts emerge from the foregoing analysis :

1. In some parts of the Iril drainage basin, the number of streams and confluences are equal to the number of tributaries which join a major stream;
2. The parts of the basin, which are highly dissected have fine drainage texture and vice versa; and
3. In the highly dissected area of radial drainage pattern between Sanakeithel and Boljang, and in the western and south - eastern part of Saikul, the number of streams per sq km is larger than the number of confluences.

The study of drainage texture and number of confluences is helpful in deciphering the influence of physiography on drainage texture. It also helps in understanding the nature of dissection.

To conclude, we may say that there is a close relationship between drainage texture and intensity of erosion. The regions of fine drainage texture suffer immensely from rapid erosion.

3.7 DRAINAGE DENSITY

Drianage density is an important indicator of the linear scale of landform elements in a drainage basin (Horton, 1945). The views of Horton (1945) and Strahler (1964) and others on drainage density are almost similar. According to the former, "drainage density is the total length of the streams in a given drainage basin divided by the area of the basin". The latter states, "in order to find out drainage density the lengths of the streams of all hierarchical orders should be measured, totalled up and divided by the geographical area of the drainage basin". Dury (1969) has used the expression." texture of dissection" and says, " in order to express the texture of dissection, we must use not the water course but the valley bottom (the thalweg), whether wet or dry". The ratio thalweg length divided by area is equivalent to drainage density as suggested by Horton and others. This may be taken as the index of dissection of a drainage basin. Two areas may have identical drainage density but while one of them may have higher dissection index because of the presence of a large number of streams, the other may have lower dissection index because of the presence of small number of streams.

Making use of the drainage map (Fig.3.1), the lengths of the streams occurring within each of the 1285 squares were measured separately and totalled up. The figures were indicated at the centres of the squares. Selecting a suitable class interval, three indices of dissection ranges, *i.e.* high, medium and low were formulated. The squares were distributed among these three categories (Table 3.3). Isopleth lines were then drawn on a map with the help of the figures indicated in the squares and a choropleth map was also prepared by shading the squares in accordance with the category to which they belonged (Figs. 3.5 and 3.5a).

Table 3.3
Drainage Density or Texture of Dissection

Sl.No.	Thalweg length in km.	Number of squares	% of the total	Category of drainage density or texture of dissection
I	Below 2	366	28.48	Low
II	2 - 4	760	59.14	Medium
III	4 - 6	159	12.38	High
		1285	100.00	

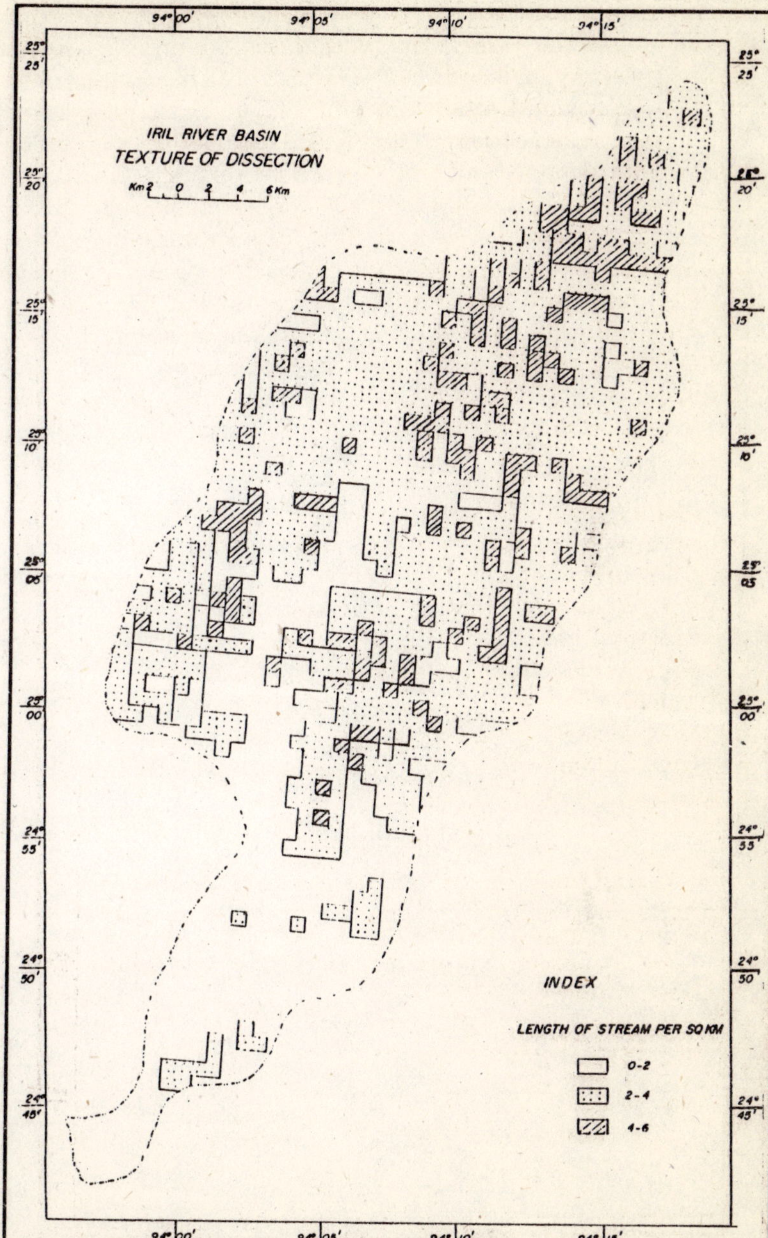
IRIL RIVER BASIN
TEXTURE OF DISSECTION
Km 2 0 2 4 6 Km
94° 00'
94° 05'
94° 10'
94° 15'
25° 25'
25° 20'
25° 15'
25° 10'
25° 05'
25° 00'
24° 55'
24° 50'
24° 45'
INDEX
LENGTH OF STREAM PER SQ KM
0-2
2-4
4-6

Fig. 3.5

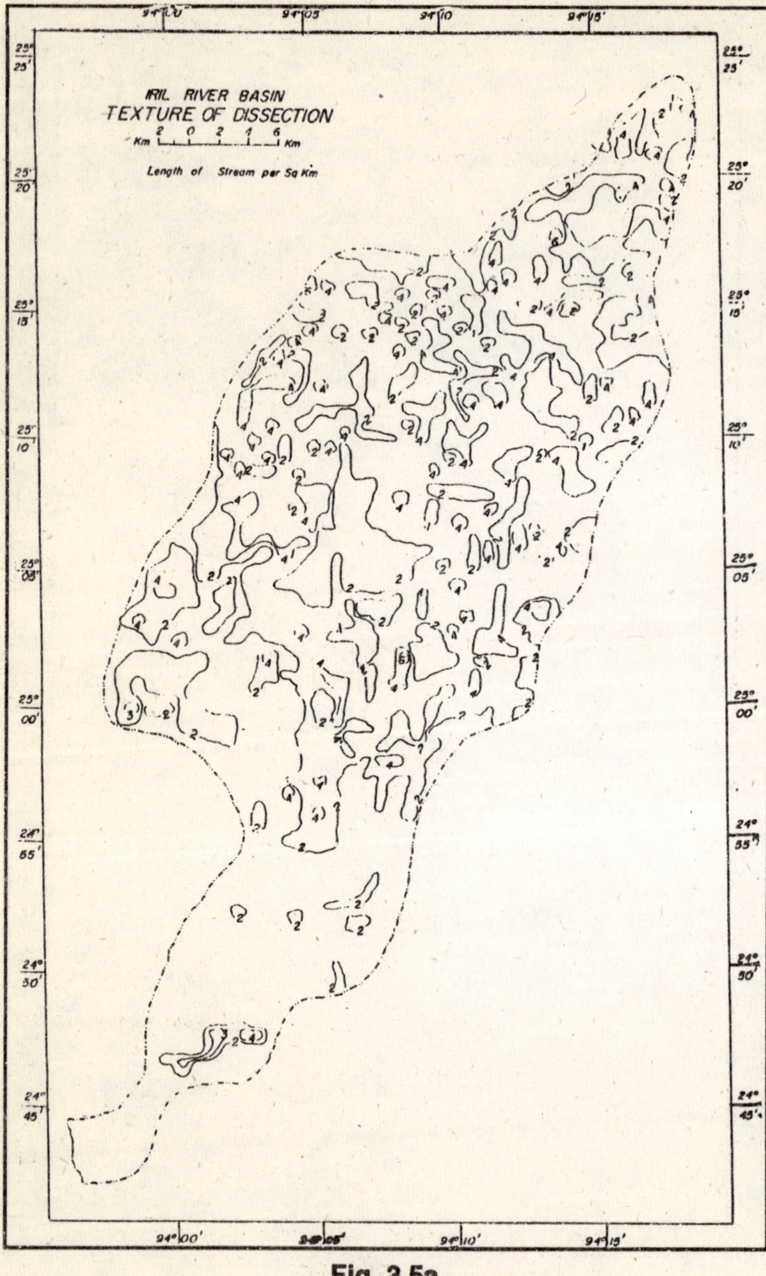
IRIL RIVER BASIN
TEXTURE OF DISSECTION
Km 2 0 2 4 6 Km
Length of Stream per Sq Km

Fig. 3.5a

Figs. 3.5 and 3.5a show the distribution of drainage density. This may be compared with the map depicting number of streams per sq km (hydrographic map). The comparison gives an idea of the distribution of texture of dissection categories in the Iril basin. A study of Figs. 3.5 and 3.5a together with Table 3.3 reveals that the highly dissected lands are found scattered, but show a tendency to agglomerate in the central and south western parts of the Iril basin. They claim about 12.38 per cent of the total area of the basin. Here the indices of dissection are high, varying between 4 and 6 km of the thalweg length. The parts of the basin with medium texture of dissection encompass the highest proportion (59.14 per cent) of its total area. The central part of the basin, comprising the valley of the Iril and the area around its mouth, where relief is insignificant, have low texture of dissection. They constitute 28.48 per cent of its total area.

3.8 CONCLUSION

It may be concluded from the preceeding analysis that the parts of the Iril river basin, which have high dissection indices, have 12 to 16 streams per sq km and 4 to 6 km of streams per km of thalweg length indicating high drainage density. Those with medium dissection indices have 4 to 12 streams per sq km and 2 to 4 km of streams per km of thalweg length showing medium drainage density. The parts of the basin which have low dissection indices have 0 to 4 streams per sq km and 0 to 2 km of streams per km of thalweg length which point towards low drainage density.

After analysing various factors of stream order, drainage pattern and its significance, number of streams, number of confluences, drainage texture and drainage density, it may be stated that the parts of the basin where both the minor and the major streams are distinct, and where the length of the minor streams exceeds that of the major streams per unit area, the land is highly dissected and the drainage texture is fine. Where the minor streams join the major streams and the stream length and stream number decrease, it may be stated that the land is less dissected and the drainage texture is coarse.

The high dissection and the fine texture of drainage were observed on the higher reaches where the valley is relatively narrow. As the river descends, the middle zone of its course is characterised by medium dissection, medium texture and medium drainage density. The lower reaches of the Iril basin show the picture of a mature river with meandering courses in its broad valleys. Therefore, it can be inferred that the Iril river exhibits a polycyclic landscape.

4

BASIN MORPHOMETRY

One of the most fundamental geomorphic units of an area is the drainage basin having its own characteristics and stages of development. Running water is one of the main agencies shaping the relief by sub - aerial process. Morphometric analysis can reveal the degree to which the present processes have externally or internally adjusted with variables of the basin in relation to each other. Therefore, the present study peeps into the behavioural pattern of drainage network and development of morpho - units showing whether they are all evolved in a similar manner in relation to each other or not.

4.1 SELECTED DRAINAGES

The Iril basin is constituted by several smaller sub - basins of different magnitudes. For the present analysis, six main tributaries of Iril river. *viz.*, Ithoi Lok, Ihang Lok, Ekou Lok, Leimakhong and Wari Lok have been taken up. All of them provide convenient units for measurement and analysis of their morphometric properties and for the proper understanding of the basin region.

4.1.1 ITHOI LOK

The Ithoi Lok or Ithoi river is the biggest left bank tributary of Iril river which rises from a height of 1860 m. It collects several smaller streamlets such as Darai Lok, Shorbung Lui, Selshi Vadung, Chaljang Lui, Shumlong Lok, Tengkon Lui, Gova Dung, Taitoi Lui, Selsi Lui etc. This drainage basin covers an area of 237.50 sq km in the north western part of the Iril basin. Flowing through hilly tracts in a zig - zag fashion, Ithoi Lok joins the Iril near Matakhong village at a level of 795 m above MSL. Many of the smaller streams do not contain water during the winter season. However, during the rainy season they carry running water to discharge in the main tributary.

4.1.2 IHANG LOK

The Ihang Lok is an important right bank tributary of the Iril river. It rises as Ngarong Khong at a height of 1860 m above MSL. After receiving its own tributaries - the Khengvei Khong (source ht. 1500m) and Chichi Khong (source ht. 1500 m), it takes a sharp bend and assumes the name Ihang Lok, which then flows more or less parallel with Iril river for about 36.5 km in NE - SW direction. The Ihang Lok then bends towards NW direction near Sanakeithel to unite with Iril river at the level of 800 m. Just after the confluence with Ihang Lok at Ihang Karong, the Iril river starts showing its youthful stage. Before its confluence with the Iril, this sub - basin covers an area of 196.25 sq km in the eastern and south - western parts of the Iril basin collecting water from its feeder streams *viz.* Langsu Kongri, Khonpang Kongri, Ara Khong, Vahna Khong, Moreang Khong etc. As the relief of the source of these streams is much higher than the confluence point, they excavate the valleys with high energy. However, field checking reveals that for the major part of the year, these smaller streams are without water. During the rainy season every small stream carries water excavating their valley courses. The Ihang river itself flows through undulating hills with broad river bed.

4.1.3 ICHAI LOK

Ichai Lok is also a left bank tributary of the Iril river flowing in the western part of the basin. It originates at a height of 1690 m near the village Jugidara and occupies some of the south - western parts of the Iril basin. It joins the Iril river at a height of 800 m near Saikul, and occupies a drainage basin of 59.02 sq km area. Its basin is characterised by hills and broad river beds.

4.1.4 EKOU LOK

Ekou Lok, another left bank tributary, covers an drainage area of 56.14 sq km in the Iril basin. Originating at a height of 1540 m in the south western part of the basin, it joins the Iril river at Ekou Bazar. The other side of the ridge, beyond the water divide, is drained by the Imphal drainage system. The surrounding hills are almost barren due to deforestation. The small valley portions along the river are used for cultivation by the nearby inhabitants.

4.1.5 LEIMAKHONG RIVER

Leimakhong river is another right bank tributary of the Iril originating from a height of 1,440 m and covering a basin area of 112.03 sq km. This basin has lesser number of streams in comparison with its area which indicates a less dissected part of the Iril basin. Most part of the drainage flows through valley area thereby having less number of streams that too with broad valleys.

4.1.6 WARI LOK

Wari Lok originates at a height of 1880 m near the Laichan village covering a small area in the north - western part of the basin. It covers an area of 34.79 sq km in the Iril basin and flows through the hilly terrain/dissected hills, thereby showing a river bed with high relief.

These smaller basins differ in their terrain and lithology. Stream channels or drainage lines are the most important characteristic features for defining a fluvial system in basins with differential terrain. These features are quantified through analysis of a number of variables from 1 : 50,000 scale toposheets. An attempt has been made to draw a comparison among these geomorphic units.

4.2 QUANTITATIVE ANALYSIS OF THE DRAINAGE BASIN

Quantitative analysis is applied to normally developed watersheds in which running water and associated mass gravity movemants are the main agents in developing surface geometry. The importance of the quantitative analysis of the drainage basins was first felt by the publication of the classic paper of Horton (1932). It includes the order, number and length of streams, length ratio, drainage density, stream gradient, length of overland flow, constant of channel maintenance, adventitious streams and form or shape of drainage basins, which have been considered for selected drainages of the Iril river basin in this section.

4.2.1 ORDER, NUMBER AND LENGTH OF STREAMS

The number of different orders of the stream segments have been counted. Among the six tributaries/basins, Ithoi basin, having an area of 237.5 sq km is found to be the biggest with the highest number of streams, *i.e.* 1,875 out of the total 4,576 streams for all the six drainage basins.

In the sample basins, Ihang, Ithoi and Leimakhong are the 6th order streams and Ekou, Ichai and Wari are the 5th order streams. Table 4.1 shows the number of streams according to different orders in different drainage basins of the Iril.

Horton (1945) established a geometrical relationship amongst stream length, stream number and stream order. According to him, the number of streams of different orders in a given drainage tend closely to approximate an inverse geometric series in which the first term is unity and the ratio is the bifurcation ratio. From Table 4.1 it is seen that the average bifurcation values of this basin conforms to the value calculated by Horton (1945) and Strahler (1957). According to them bifurcation ratios characterstically range between 3.0 to 5.0 for watersheds in which the geologic structures do not distort the drainage pattern. However, in this region, the average value of each of the streams does not exceed the value calculated by Horton and Strahler, one single value for Ihang lok is as higher as 7.0 with the minimum possible value of 2. Other tributaries, such as Ekou and Leimakhong, also touch the minimum value of 2. This explains that the area of Ithoi basin is somewhat distorted by joints and faults in some portions of the basin. Again the relations of number of streams of each order to the next higher number have been shown in Figs. 4.1 and 4.2. The graph would have shown a marked - up concavity at the lower end if all the points were joined as it has been done by Horton (1945). This indicates that geometric progression is not closely observed in the higher orders (Schumm, 1956). Further, from Table 4.1 it is observed that with the increase in order the number of stream decreases and upholds Horton's law.

By studying stream frequency of the basin a perfect inter - relation between stream order and drainage texture may be established. Actually stream frequency indicates the number of streams per unit of area (Horton, 1945). The values of stream frequency per sq km are high within the lower stream order, which indicates the zone of fine texture, and the lower frequency value within the higher order indicates the approaching maturity of the basin development. The higher stream frequency value in case of Ihang and Ithoi indicates a higher and more dissected terrain than the other basins. The higher frequency value in the first order of Leimakhong is due to small area in the first order. In case of Ekou, the frequency value for first order is much lesser than that of others which shows a maturity stage in the higher order. Although the basin area of Wari Lok is small, the stream

Table 4.1
Number of Streams, Bifurcation ratio and Stream Frequency

Stream Order	Number of Streams						Bifurcation Ratio						Stream Frequency					
	Ihang	Ithoi	Leima-khong	Ekou	Ichai	Wari	Ihang	Ithoi	Leima-khong	Ekou	Ichai	Wari	Ihang	Ithoi	Leima-khong	Ekou	Ichai	Wari
1	1119	1478	206	214	350	193							37.66	37.62	40.47	11.37	31.78	28.05
2	246	306	48	52	67	36	4.54	4.83	4.29	4.11	5.2	5.36	4.40	4.53	2.40	1.89	5.73	3.15
3	55	70	10	11	16	11	4.4	4.37	4.8	4.72	4.18	3.27	0.82	0.96	0.37	0.33	1.01	0.09
4	15	16	5	2	3	3	3.9	3.68	3.33	5.5	5.3	3.6	0.13	0.10	0.18	0.07	0.07	0.12
5	2	4	2	1	1	1	7	4.75	1.5	2	3	3	0.016	0.02	0.03	0.01	0.01	0.02
6	1	1	1				2	4	2				0.005	0.004	0.008			
Total Average	1438	1857	272	280	437	244	4.36	4.32	3.18	4.08	4.42	3.82						

Table 4.2
Components of Stream Length

Stream Order	Total Length (in km)						Mean Length (in km)						Cumulative Mean Length					
	Ihang	Ithoi	Leima-khong	Ekou	Ichai	Wari	Ihang	Ithoi	Leima-khong	Ekou	Ichai	Wari	Ihang	Ithoi	Leima-khong	Ekou	Ichai	Wari
1	383.35	569.36	113.15	94.15	84.75	145.82	0.34	0.38	0.54	0.43	0.24	0.75	0.34	0.38	0.54	0.43	0.24	0.75
2	120.03	177.51	35.92	25.25	28.50	22.26	0.48	0.58	0.74	0.48	0.42	0.61	0.82	0.96	1.29	0.92	0.66	1.37
3	68.09	66.76	25.82	16.05	12.07	10.37	1.23	0.95	2.58	1.45	0.75	0.94	2.05	1.91	3.87	2.38	1.42	2.31
4	29.85	45.10	12.70	7.25	11.1	7.20	1.99	2.81	3.17	3.62	3.70	2.40	4.04	4.72	7.05	6.00	5.12	4.71
5	25.4	27.7	14.5	6.8	7.25	4.75	12.7	6.92	7.25	6.8	7.25	4.75	16.74	11.64	14.30	12.80	12.37	9.46
6	20.7	24.5	7.00				20.7	24.5	7.00				37.44	36.14	21.30			

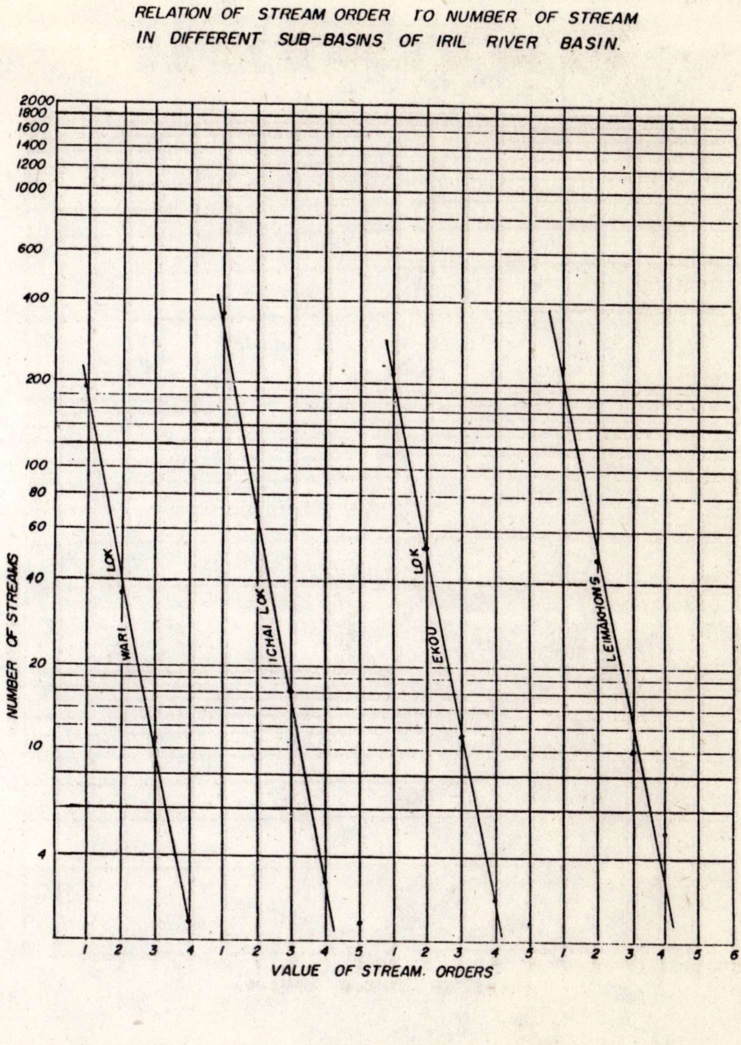
RELATION OF STREAM ORDER TO NUMBER OF STREAM
IN DIFFERENT SUB-BASINS OF IRIL RIVER BASIN.
2000
1800
1600
1400
1200
1000
600
400
200
100
80
60
40
20
10
4
NUMBER OF STREAMS
WARI LOK
ICHAI LOK
EKOU LOK
LEIMAKHONG
1 2 3 4 1 2 3 4 5 1 2 3 4 5 1 2 3 4 5 6
VALUE OF STREAM. ORDERS

Fig. 4.1

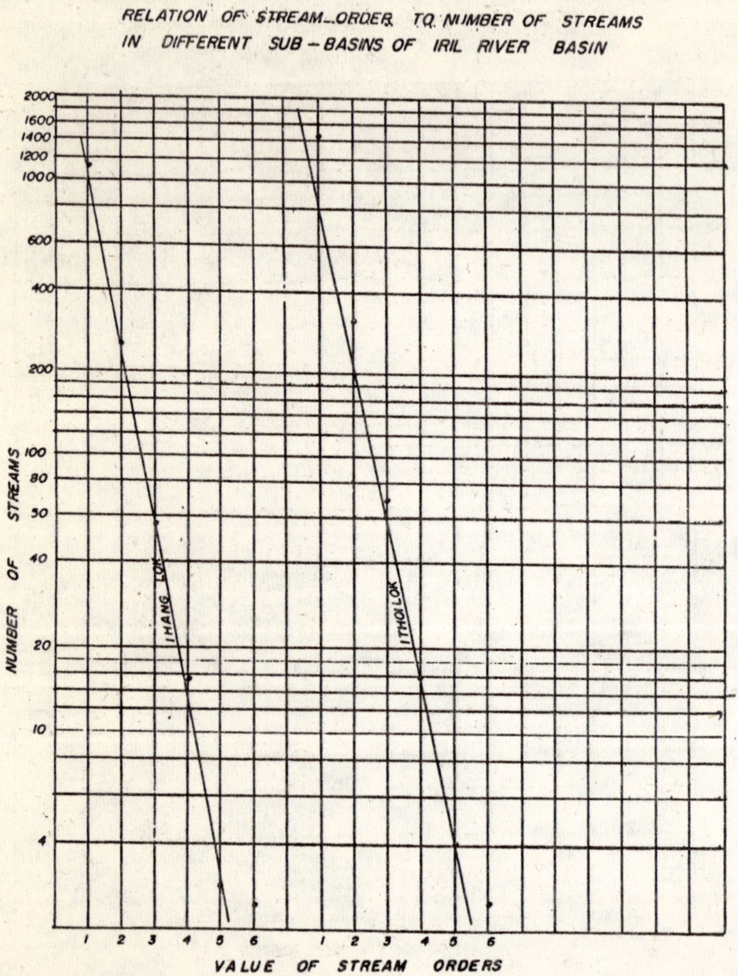
RELATION OF STREAM-ORDER TO NUMBER OF STREAMS
IN DIFFERENT SUB-BASINS OF IRIL RIVER BASIN
NUMBER OF STREAMS
2000
1600
1400
1200
1000
600
400
200
100
80
50
40
20
10
4
IHANG LOK
ITHOILOK
1 2 3 4 5 6
1 2 3 4 5 6
VALUE OF STREAM ORDERS

Fig. 4.2

frequency is much higher in the first order and less mature in the advance stage, thus indicating higher scarp with dissected terrain. This can be confirmed from the Table 4.4 where a general frequency value for the selected drainage basins are given. In Table 4.5, Leimakhong shows the lowest value of stream frequency, thus indicating mature and advance development of the basin in the higher stream order. Among the six selected basins, Leimakhong is the most matured drainage followed by Ekou Lok.

The study of length of streams, with its mean values in relation to order number, is of great importance in the study of the drainage characteristics of a basin. As the stream order goes higher, the length of stream also becomes higher. The cumulative mean length also shows the same pattern of progressive increase with increase in order (Table 4.2). The lower value of mean length for the first order stream shows that most of the first order streams are short and, therefore, expresses more dissected area. Although the Wari Lok is smaller in comparison to other rivers, its mean length value is quite high. A comparison of the two tables *i.e.* Table 4.1 and 4.2 indicate that though Ithoi has the highest number of first order streams, its mean length value is not low, and Ihang has less mean length value for the first order stream although it has higher number of first order streams.

4.2.2. STREAM LENGTH RATIO

The ratio of average length of streams of a given order to that of streams of the lower order is known as the stream length ratio (Satpathi, 1981). The length ratio should range between 2 and 3 as estimated by Horton (1945).

Table 4.3
Stream Length Ratio (r_1)

Tributary	l_2/l_1	l_3/l_2	l_4/l_3	l_5/l_4	l_6/l_5
Ihang	1.41	2.56	1.61	6.36	1.62
Ithoi	1.52	1.63	2.94	2.46	3.54
Leimakhong	1.37	3.48	1.22	2.28	0.95
Ekou	1.11	3.02	2.49	1.88	
Ichai	1.75	1.78	4.93	1.95	
Wari	0.81	1.54	2.55	1.97	

Where r_1 = mean length ratio

Table 4.4
Areal Components of Drainage Basins

Stream Order	Total Basin Area (in sq. km)						Mean Basin Area						Percentage to the total Basin Area					
	Ihang	Ithoi	Leima-khong	Ekou	Ichai	Wari	Ihang	Ithoi	Leima-khong	Ekou	Ichai	Wari	Ihang	Ithoi	Leima-khong	Ekou	Ichai	Wari
1	29.71	39.28	5.09	18.82	11.01	6.88	0.026	0.02	0.02	0.08	0.031	0.035	15.13	16.53	5.19	34.37	19.26	18.44
2	55.79	67.51	19.95	27.48	11.69	11.40	0.226	0.22	0.29	0.52	0.174	0.316	28.42	28.42	20.35	50.19	20.45	30.64
3	66.57	72.45	26.98	32.85	15.82	11.80	1.21	1.03	2.69	2.98	0.98	1.08	33.92	30.50	27.53	60.00	26.76	31.72
4	108.27	146.28	57.67	26.53	38.68	24.86	7.21	9.14	11.53	13.26	12.87	8.28	55.16	61.59	58.84	48.48	67.68	71.08
5	119.95	174.59	63.11	56.14	59.02	34.97	59.97	43.64	31.55	56.14	59.02	34.97	61.12	73.51	56.33	100	100	100
6	196.25	237.50	112.03				196.25	237.50	112.03				100	100	100			

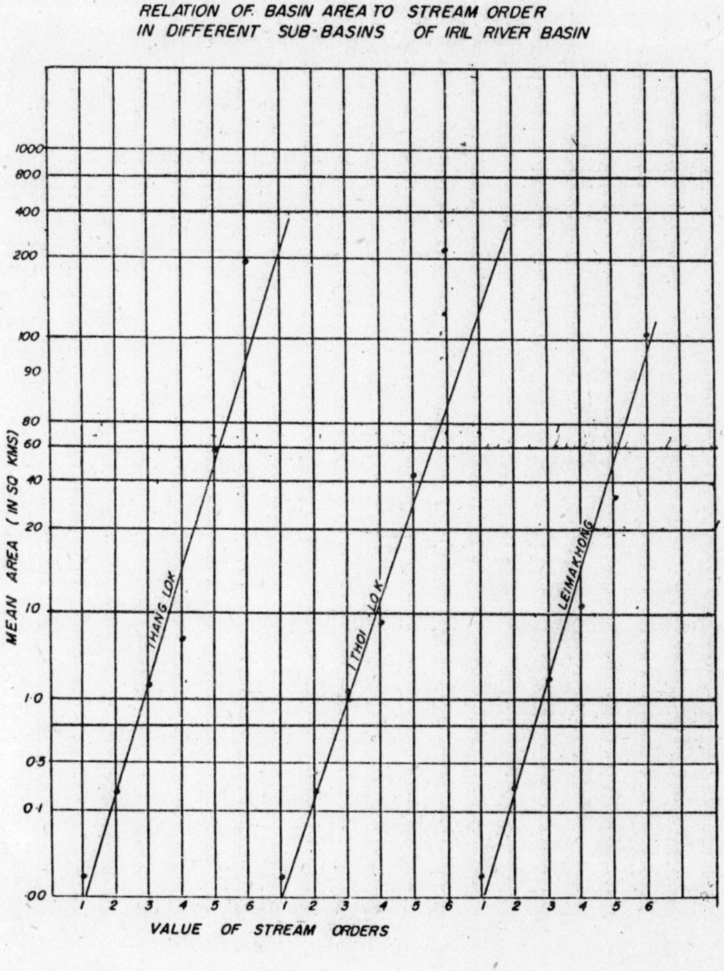
RELATION OF BASIN AREA TO STREAM ORDER
IN DIFFERENT SUB-BASINS OF IRIL RIVER BASIN
MEAN AREA (IN SQ KMS)
1000
800
400
200
100
90
80
60
40
20
10
1·0
0·5
0·1
·00
THANG LOK
ITHOI LOK
LEIMAKHONG
1 2 3 4 5 6 1 2 3 4 5 6 1 2 3 4 5 6
VALUE OF STREAM ORDERS

Fig. 4.3

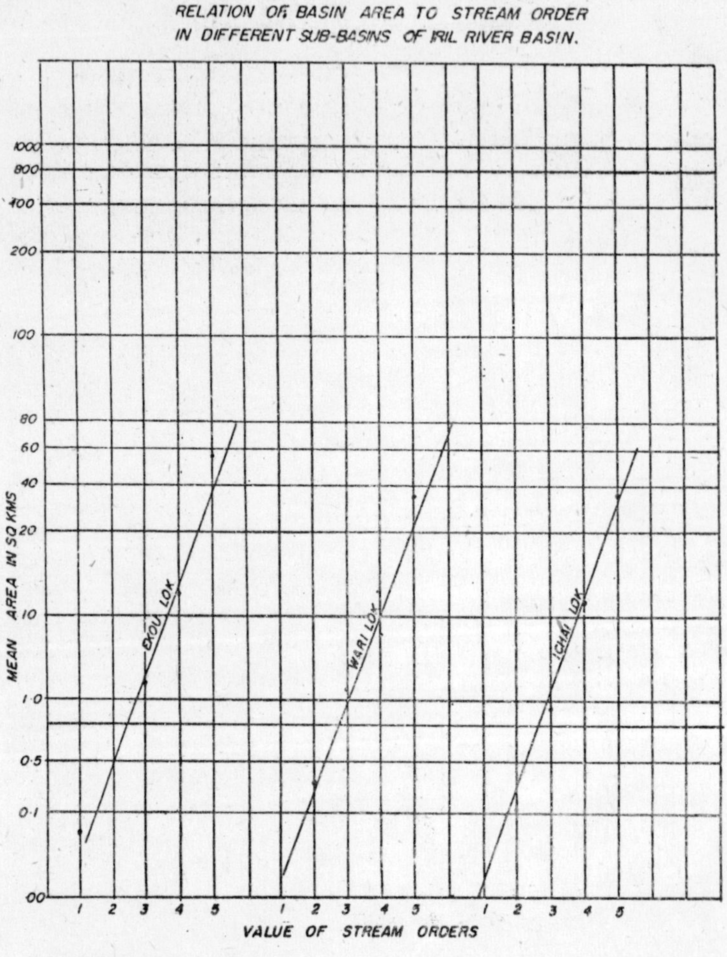
RELATION OF BASIN AREA TO STREAM ORDER
IN DIFFERENT SUB-BASINS OF IRIL RIVER BASIN.
1000
800
400
200
100
80
60
40
20
10
1.0
0.5
0.1
00
MEAN AREA IN SQ KMS
EKOU LOK
WARI LOK
ICHAI LOK
1 2 3 4 5
1 2 3 4 5
1 2 3 4 5
VALUE OF STREAM ORDERS

Fig. 4.4

However, the length ratios of the drainage basins of Iril river do not totally conform to that of Horton (Table 4.3). The number of streams is dependent on the bifurcation ratio and increase with stream order while the length per stream is dependent on the stream length ratio (r_1) and decrease with increasing stream order.

According to Horton (1945), the mean drainage area should form a geometric series. Again, the mean stream length and mean area of the drainage basins have a geometric relationship with each order. Table 4.4 shows the mean basin area for the six selected drainage basins and Figs. 4.3 and 4.4 reveal a close geometric relationship between the stream order and the average drainage basin area of each order.

The relationships between the basin area and the stream length have been observed and presented in Table 4.5. The relationships between stream length, drainage basin area and the stream order number are dependent on the Constant of Channel Maintenance. "The relationships between mean drainage basin area of each order and mean channel lengths of each order of any drainage network is a linear function whose slope (regression coefficient) is equivalent to the area in square feet necessary on the average for the maintenance of one foot of drainage channel" (Schumm, 1956).

Table 4.5
Drainage Basin Characteristics

Name of the Basin	Stream Frequency (F_s)	Drainage Density (D_d)	Length of Overland Flow (I_o)	Constant of Channel Maintenance (C)
Ihang	7.32	3.3	0.15	0.30
Ithoi	7.89	3.8	1.31	
Leimakhong	2.42	2.13	0.23	0.46
Ekou	5.11	2.73	0.18	0.36
Ichai	7.64	2.51	0.19	0.39
Wari	6.55	5.11	0.09	0.19

$I_0 = 1/2\,D_d$ $\quad$ $C = a/\ L = 1/D_d$

4.2.3 DRAINAGE DENSITY

Among the six drainage basins, Wari basin shows the highest drainage density followed by Ithoi and the Ihang. The Leimakhong shows the lowest drainage density among them as it runs almost over the plain without much gradient in the upper reaches and thus showing low gradient value, which remains almost unchanged in the third and fourth orders and again at the fifth and sixth orders (Table 4.5), whereas the streams of high drainage density such as Wari, Ithoi and Ihang drain the area of high relief and receive higher amount of rainfall etc. The basins Ekou and Ichai are lower than Wari, Ihang and Ithoi, thus indicating somewhat lower gradients. Rock formation and infiltration capacity of the bed rock also affect the drainage density.

4.2.4 MEAN STREAM GRADIENT

The average drop in elevation from the head waters to the mouth of all stream segments of a given order divided by the average length of those segments is termed as the mean stream gradient (Ball, 1881). The gradients progressively decrease from the lower order to the higher order indicating the highest gradient value in the first order as shown at Table 4.6. There is a fairly definite relationship between the slope of the streams and stream order which can be expressed by an inverse geometric series law (Schumm, 1956).

Table 4.6
Mean Gradients of Drainage Basins

Drainage Basin	Mean Gradient of 1st order Stream	Mean Gradient of 2nd order Stream	Mean Gradient of 3rd order Stream	Mean Gradient of 4th order Stream	Mean Gradient of 5th order Stream	Mean Gradient of 6th order Stream
Ihang	2.91	2.01	0.80	0.50	0.078	0.048
Ithoi	2.59	2.60	1.04	0.35	0.14	0.04
Leimakhong	1.82	1.33	0.38	0.39	0.14	0.14
Ekou	2.27	2.05	0.68	0.27	0.14	
Ichai	4.15	2.35	1.32	0.27	0.13	
Wari	1.32	1.61	1.06	0.41	0.21	

4.2.5 LENGTH OF OVERLAND FLOW

One of the important factors that help us to know the amount of water needed to exceed erosion is length of overland flow. This is the distance between two adjacent channels which is reciprocal of their drainage density. One half of this distance is known as the length of overland flow and is synonymous with length of sheet flow. This is highly responsible for soil erosion which is more prominently seen along the unterraced hilly slopes of this region. Length of overland flow value at Table 4.5 shows that the length of flow is the highest in the Ithoi due to its flatness in valley portion, whereas due to steeper slopes and highly undulated nature of the landscapes the length of overland flow is much lower in Wari and other basins *i.e.* Leimakhong, Ihang, Ekou and Ichai. For determining various characteristics such as hydrological nature, physiographical character and erodibility of a drainage basin, length of overland flow is one of the important independent variables influencing and affecting them.

4.2.6 CONSTANT OF CHANNEL MAINTENANCE

For observing the nature of run - off, surface erosion and drainage network development, the measurement of the constant of channel maintenance is highly required. Along with drainage density, this constant is of value as a means of comparing the surface erodibility or other factors affecting surface erosion and drainage network development (Schumm, 1956). The value as shown at Table 4.5 for different drainage basins indicate that Wari, Ihang and Ithoi require the minimum area for the development of drainage chennel whereas the Liamakhong needs maximum area for its drainage chennel development.

4.2.7 ADVENTITIOUS STREAMS

According to Horton (1945), adventitious streams are results from accidental variation of conditions within area, where some of the first and second order streams enter the main stream directly and not through some other higher order. This type of stream increases the number of streams of lower orders, thus increasing the bifurcation ratio greater than 2.0, which is common in the natural streams. From Table 4.7 it is seen that Ihang and Ithoi, being biggest basins, have the highest number of adventitious streams followed by Ichai, Wari, Ekou and

Leimakhong. Adventitious streams are generally developed at the later stage of a river, *i.e.* at the maturity stage.

Table 4.7
Adventitious Streams in the Drainage Basins

Name of the Basin	Adventitious Stream		Total
	1st Order	2nd Order	
Ihang	66	22	88
Ithoi	71	11	82
Leimakhong	8	4	12
Ichai	37	11	48
Ekou	8	10	18
Wari	29	10	39

4.2.8 FORM OR SHAPE OF DRAINAGE BASINS

In the study of a basin, the analysis of its shape is indispensable. A fan - shaped catchment produces greater flood intensity as it has less concentration than a fern - shaped catchment as the time of concentration is more and the discharge is distributed over a long period (Raghunath, 1986). Generally, the shape of a drainage basin can be expressed by the form factor, compactness coefficient, elongation ratio and circularity ratio (Table 4.8).

a. Form Factor (F_f)

Horton (1932) made a quantitative expression of drainage basin outline form through a form factor, which is the dimensionless ratio of basin area (A) to the square of basin length (L_b), thus the formula A/L_b^2 is found out.The ratio, found after inverting it L_b^2/A is used in the study of hydrograph (form factor).

Table 4.8
Shape of Drainage Basins

Name of the Basin	Form Factor (F_f)	Compactness Coefficient (C_c)	Elongation Ratio (E_r)	Circularity Ratio (C_r)
Ihang	6	1.61	0.433	0.383
Ithoi	4	1.26	0.08	0.627
Leimakhong	6	1.71	0.426	0.338
Ekou	2	1.13	0.602	0.785
Ichai	3	1.23	0.666	0.663
Wari	2	1.24	0.291	0.652

b. Compactness Coefficient (C_c)

This is another ratio which is independent of the size of the catchment and is dependent only on the slope of the basin area.

c. Elongation Ratio (E_r)

Schumm (1956) used an elongation ratio to describe the shape of a basin which varies due to differences in relief, climatic condition and geological structure. This ratio is defined as the ratio of the diameter of a circle of the same area as the basin to the maximum basin length. Generally in strong relief, the value ranges from 0.6 to 0.8. As the study area is in higher relief with minimum valley portion, except Leimakhong all drainage basins show higher elongation ratio which varies from 0.2 to 0.6, thus indicating that the basin occupies both the higher relief with medium and low relief regions.

d. Circularity Ratio (C_r)

Miller (1953) used circularity ratio, which is defined as the ratio of the basin area to the area of a circle having the same perimeter as the basin. This ratio varies with the shape of the basin. A much elongated basin shape produces somewhat higher circularity ratio. Most of the sample basins, except Ihang and Leimakhong, also show higher circularity ratio.

4.3 CONCLUSION

From the above analysis of the selected drainage sub - basins, it is seen that two of the big tributaries show higher frequency values in their lower orders indicating a well dissected terrain in their upper reaches. Lower frequency value in the higher order also suggests an advanced stage of the drainage or the mature stage. The study of drainage density also supports the above facts showing high drainage density value in the well dissected area and low value at the lower gradient or the lower reaches. From the study of shape or form of the drainages it is seen that Ihang and Leimakhong drainages have concentration of water within their basin area for a longer duration.

After an overall examination of the sub - basins, it can be concluded that Iril drainage basin is composed of various sub - basins having differential morphometric characters and erosional surfaces, which have been developed under variety of conditions. Thus, this basin shows a polycyclic drainage character of its own.

5

SLOPE AND MORPHOLOGY OF THE LANDSCAPE

The study of slope provides a lucid account of various topographical features and evidences, which help in the interpretation of the complex form of landscape. Slope also expresses most of the morphological characters of the area such as its appearance, dimension and magnitude, It is influenced by the processes of erosion, weathering, types of drainage in that particular area and climatic condition. The Iril river basin presents erosional surfaces with weathered faces, thus showing subdued topography with varying factors such as intensity of relief, relative relief, density of texture and climate. The study of contour can give a first hand impression about the general ground slope. An attempt to the study of slope and its relation to the relief has been made to find out the present morphology of the Iril river basin with the help of various methods of relief study.

In addition to the analysis of slope, other methods such as altimetric frequency histogram, hypsographic curve, area height diagram, streamline surfaces, generalised contour, cross valley profile, projected, composite and superimposed profiles of the landscape are also included in this section to portray a clear picture of the present morphology of the Iril basin landscape.

5.1 INTENSITY OF RELIEF

Average relief in different categories expresses the intensity of relief for that area. The intensity map of this basin has been prepared by finding out the average relief per unit area, *i.e.* average relief per sq km. Table 5.1 shows the distribution of intensity of average relief in different categories. The choropleth map (Fig. 5.1) shows the intensity of relief per sq km. The isopleth map (Fig. 5.1a) indicates the lowest intensity in the central and lower parts of the drainage basin. These maps define eight categories of

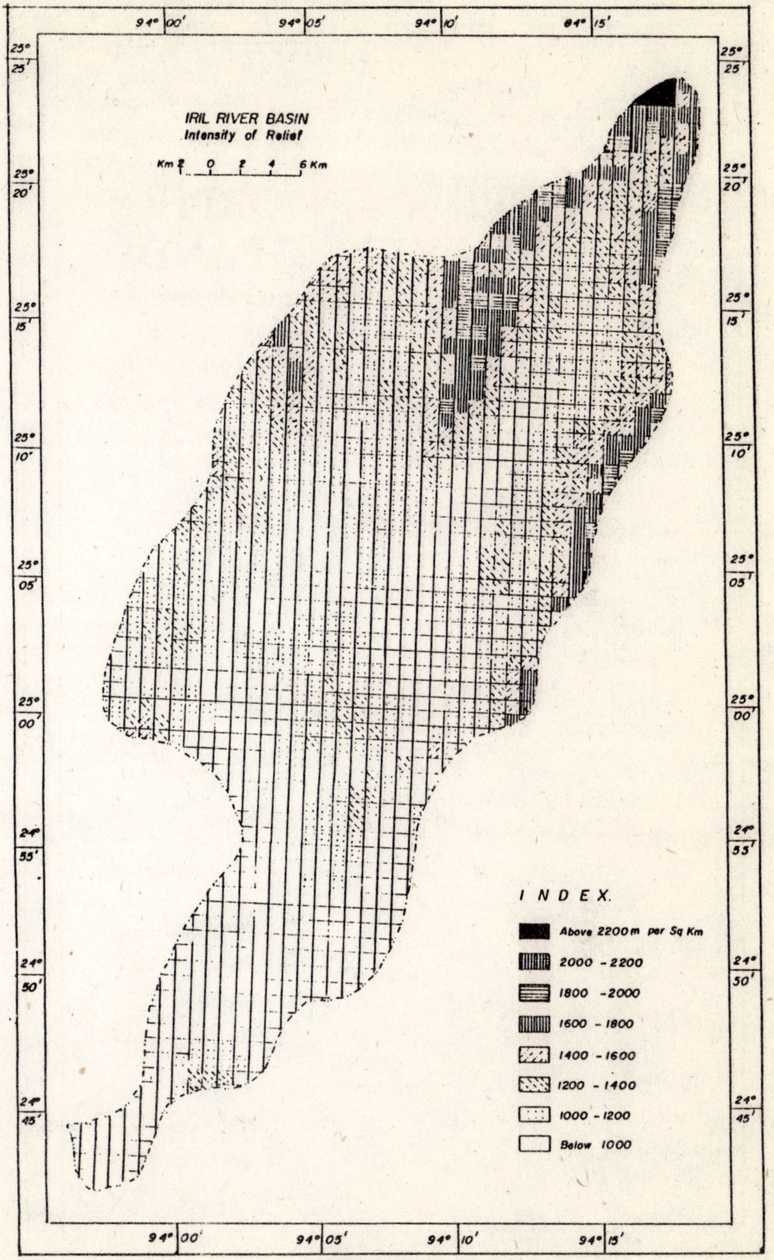
IRIL RIVER BASIN
Intensity of Relief
Km 2 0 2 4 6 Km
INDEX.
Above 2200 m per Sq Km
2000 - 2200
1800 - 2000
1600 - 1800
1400 - 1600
1200 - 1400
1000 - 1200
Below 1000
94° 00'
94° 05'
94° 10'
94° 15'
25° 25'
25° 20'
25° 15'
25° 10'
25° 05'
25° 00'
24° 55'
24° 50'
24° 45'

Fig. 5.1

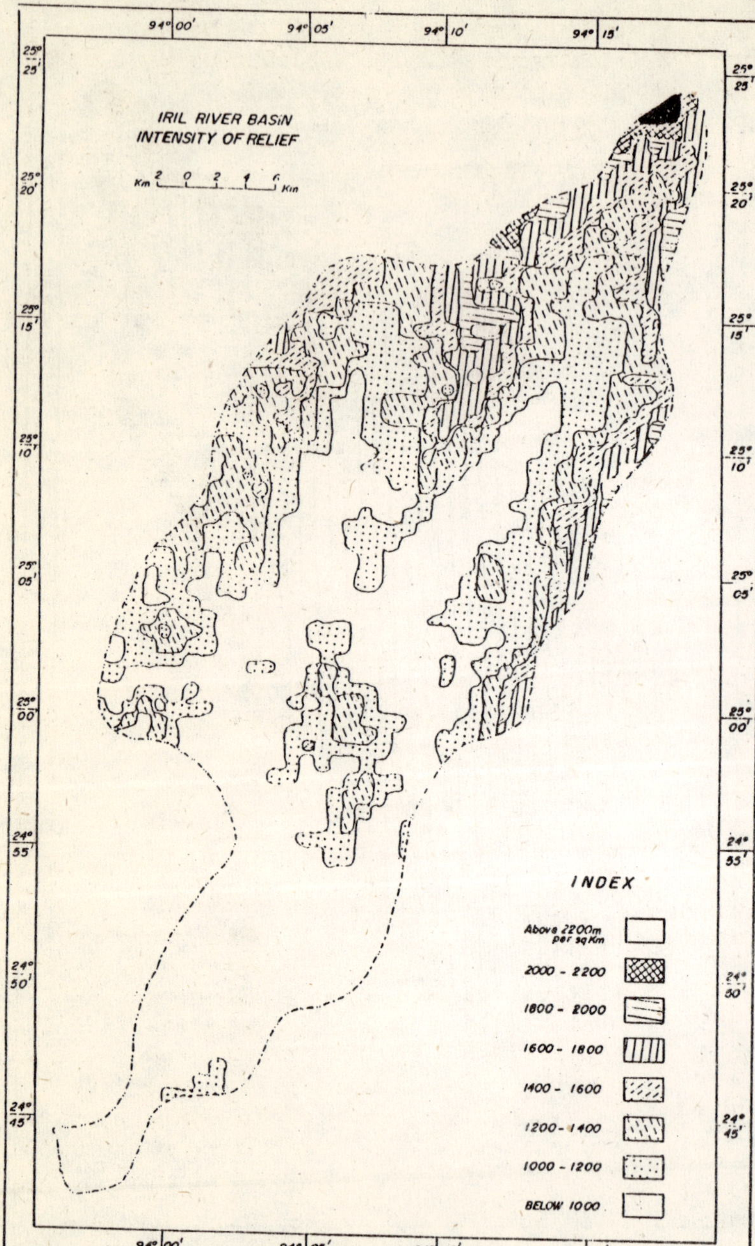
IRIL RIVER BASIN
INTENSITY OF RELIEF
Km 2 0 2 4 6 Km
94° 00'
94° 05'
94° 10'
94° 15'
25° 25'
25° 20'
25° 15'
25° 10'
25° 05'
25° 00'
24° 55'
24° 50'
24° 45'
INDEX
Above 2200m per sq Km
2000 - 2200
1800 - 2000
1600 - 1800
1400 - 1600
1200 - 1400
1000 - 1200
BELOW 1000

Fig. 5.1a

intensity of relief which can be grouped into three major categories showing 9.73 per cent as high 55.25 per cent as moderate and 35.02 per cent as low intensity zones for the entire region.

TABLE 5.1
INTENSITY OF RELIEF

Intensity of Relief (m)	Total area (sq km)	Frequency (p.c.)	Cumulative Frequency	Major categories of Intensity
Below 1000	450	35.02	35.02	Low 35.02
1000 - 1200	399	31.05	66.07	
1200 - 1400	202	15.72	81.79	
1400 - 1600	109	8.48	90.27	Medium 55.25
1600 - 1800	76	5.92	96.19	
1800 - 2000	35	2.73	98.92	
2000 - 2200	11	0.85	99.77	
Above - 2200	3	0.23	100.00	High 9.73

The high intensity zone covers a small area in the northern part of the basin while the medium or the moderate intensity is almost scattered around the central portion of the basin encircling the low intensity area in the middle of the basin extending upto the mouth of the Iril river.

5.2 RELATIVE RELIEF

The relative relief map of the area expresses the amount of slope and nature of surface slope. Following Smith's method of slope analysis (1935) the relative relief of Iril basin has been determined and mapped. The choropleth (Fig. 5.2) and isopleth (Fig. 5.2a) maps for relative relief depict the nature of surface slope per unit area and the trend of the slope respectively.

Three categories such as high, moderate and low relative relief representing an area of 23.08 per cent, 61.51 per cent and 15.41 per cent respectively can be classified (Table 5.2) and named as :

a. High relief region *i.e.* 300-500 metre and above

b. Moderate relief region *i.e.* 100-300 metre

c. Low relief region *i.e* below 100 metre.

The surface having relative relief ranging from 300 to 500 metre and above covers 23.08 per cent of the total area mostly in northern and north eastern parts of the basin. High relief regions are mostly confined to the dissected and highly undulating tracts.

TABLE 5.2
Relative Relief

Relative Relief (m)	Total area (sq km)	Frequency (p.c.)	Cummulative Frequency	Major categories of Relative Relief
Below 100	198.00	15.41	15.41	Low 15.41 %
100 - 200	531.25	41.34	56.75	Medium
200 - 300	259.20	20.17	76.92	61.51 %
300 - 400	172.25	13.41	90.33	
400 - 500	120.30	9.36	99.69	High
Above 500	4.00	0.31	100.00	23.08 %

The next important relative relief region is between 100m and 300m covering over three fifths of the total area (61.51 per cent). Moderate or medium relief regions are scattered in the western, central and southern parts of the basin. The villages, such as Mapao Tangkhul and Yasmongjong in the southern part are situated over the areas of moderate relief.

The lowest relief region occupies the southern and central sections of the basin covering around 15.41 per cent of the total area. These low relief regions are dotted in the central part distinctly marked around Ihang Karong, Sagolmang, Ekou Bazar and Pukhao area.

5.3 ABSOLUTE RELIEF

For the precise description of specific relief and for getting the direct or indirect evidences of the genesis and evolution of certain landform, absolute altitude is an important tool (Satpathi, 1981). Eight different relief regions for absolute relief are shown in Table 5.3, which reveals the following :

a. The lowest relief region, in the height range below 1000m,

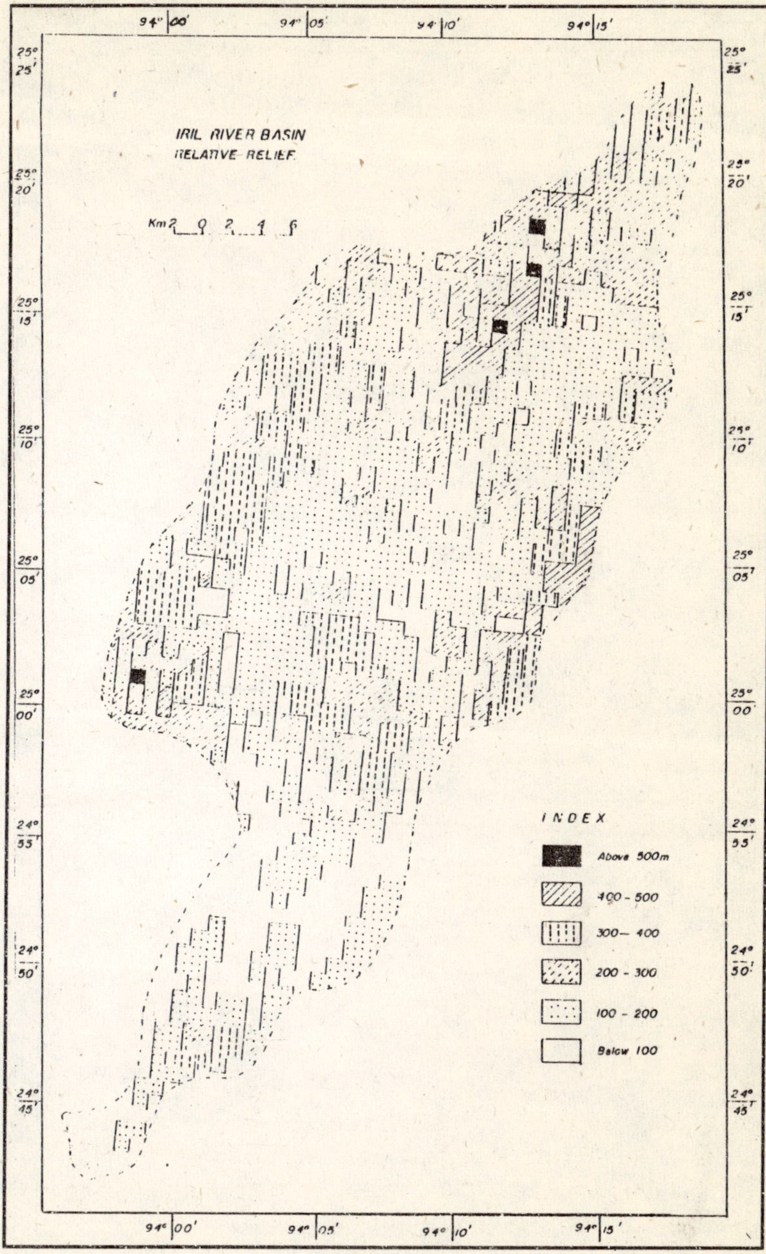
IRIL RIVER BASIN
RELATIVE RELIEF
INDEX
Above 500m
400 - 500
300 - 400
200 - 300
100 - 200
Below 100

Fig. 5.2

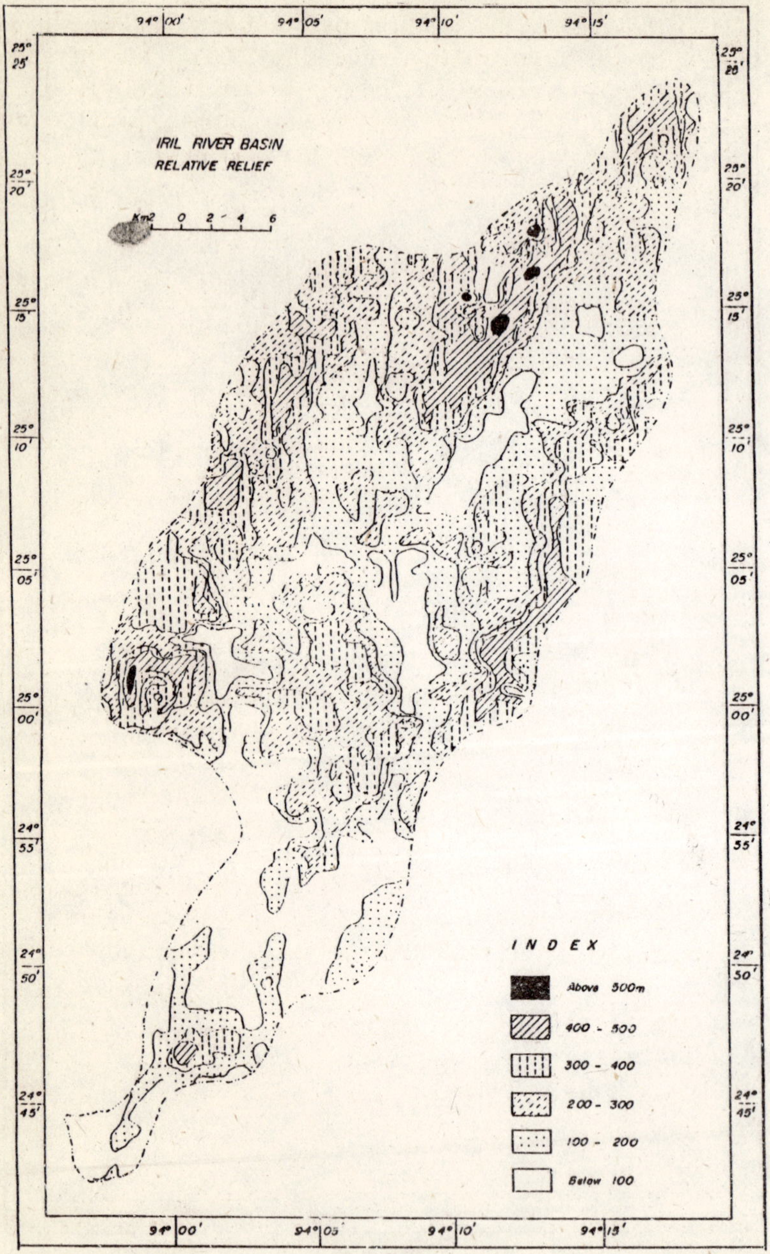
IRIL RIVER BASIN
RELATIVE RELIEF
Km2 0 2 4 6
INDEX
Above 500m
400 - 500
300 - 400
200 - 300
100 - 200
Below 100
94° 00'
94° 05'
94° 10'
94° 15'
25° 25'
25° 20'
25° 15'
25° 10'
25° 05'
25° 00'
24° 55'
24° 50'
24° 45'

Fig. 5.2a

covers most of the southern part and as some big patches within the central portion covered by 1000 m - 1200 m range. This region is almost flat covering an area of 23.91 per cent of the basin. Villages like Sagolmang, Pukhao, Lamboikhul, Taretkhul, Nongda, Kangla Siphai, Naharup etc., are situated within this height range.

TABLE 5.3
Absolute Relief

Absolute Relief (m)	Total area (sq km)	Frequency (p.c.)	cummulative Frequency	Major categories of Absolute Relief
Below 1000	307.22	23.91	23.91	Low 23.91%
1000 - 1200	497.16	38.69	62.60	
1200 - 1400	221.15	17.21	79.81	Medium
1400 - 1600	119.53	9.30	89.11	65.20%
1600 - 1800	68.16	5.31	94.42	
1800 - 2000	49.23	3.83	98.25	
2000 - 2200	17.22	1.34	99.59	High
Above - 2200	5.23	0.41	100.00	10.89%

b. The area of absolute relief having the height range of 1000m - 1200m has the largest extension in the Iril basin. It covers a large area around Shongben, Phaikon and Matakhong villages along the Ithoi river; Mollen and Sanakeithel along the Ihang Lok and Bunglung Makeng, Saikul etc, along the Ithói and Iril rivers. It covers about 38.69 per cent of the total basín area.

c. The next important region lies within a height range of 1200m - 1400m, associated with high slope regions. As it is associated with highlands, human population is much less in this region. However, some of the villages such as Washangphung Naga, Nurathel, Mongbung, Shorbung Naga, New Ngaton, Old Ngaton etc, are situated in this region.

d. The highest altitude *i.e.* above 2200m with only 0.41 per cent of the total basin area is associated with high escarpments and dense forests. The areas in the height range 1600m - 2200m

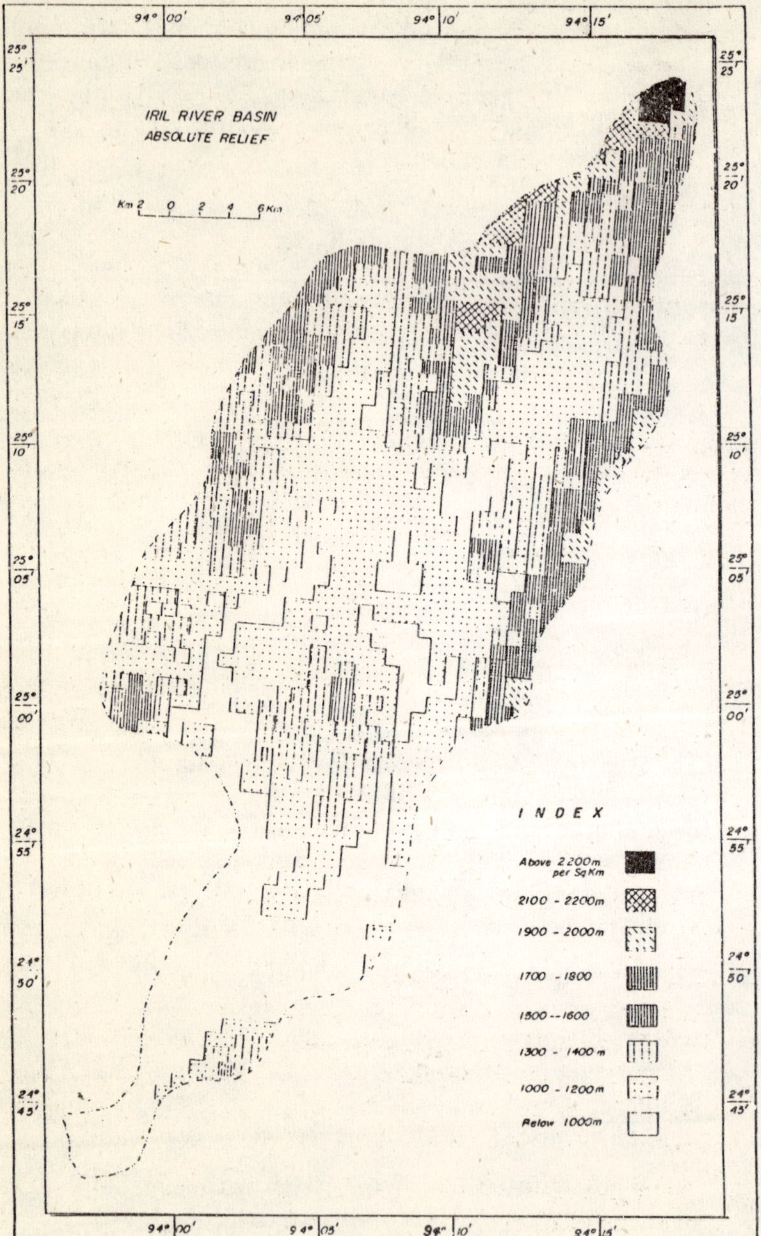
IRIL RIVER BASIN
ABSOLUTE RELIEF
Km 2 0 2 4 6 Km
94° 00'
94° 05'
94° 10'
94° 15'
25° 25'
25° 20'
25° 15'
25° 10'
25° 05'
25° 00'
24° 55'
24° 50'
24° 45'
INDEX
Above 2200m per Sq Km
2100 - 2200m
1900 - 2000m
1700 - 1800
1500 -- 1600
1300 - 1400 m
1000 - 1200m
Below 1000m

Fig. 5.3

are also associated with forests having least of human habitation in such regions.

e. The areas between 1000m and 1600m, defined as medium highland area, are associated with settlements and cover around two thirds (65.20 per cent) of the land surface of the basin. The height altitude area including the highest *i.e.* above 2200m covers an area of 10.89 per cent of the total basin.

Fig. 5.3 portrays the relief categories discussed above.

5.4 DISSECTION INDEX

The ratio between relative relief and absolute relief expresses dissection index, which varies from 0 to 0.45 in this basin showing six different ranks of index. On the basis of these six categories choropleth (Fig. 5.4) and isopleth (Fig. 5.4a) maps have been prepared. The dissection index (Table 5.4) of Iril river basin helps in estimating the vertical balance of erosion.

The table and the maps referred to clearly show the following facts :

a. The area of low relief region is associated with low dissection index. Low index of dissection covers 16.67 per cent of the total basin area which corresponds with valley section of the basin. As the low index of dissection generally corresponds with subdued relief, big patches of such areas are also scattered in the central portion of the basin while small patches are scattered randomly.

Table 5.4
Dissection Index

Index group	Total (sq km)	Frequency (p.c.)	Cumulative Frequency	Rank of Index
.00 - .05	214.20	16.67	16.67	Low 16.67%
.05 - .10	218.32	16.99	33.66	Medium
.10 - .20	484.10	37.67	71.33	54.66%
.20 - .30	310.23	24.14	95.47	
.30 - .40	52.15	4.07	94.54	High
Above - .40	6.00	0.46	100.00	28.67%

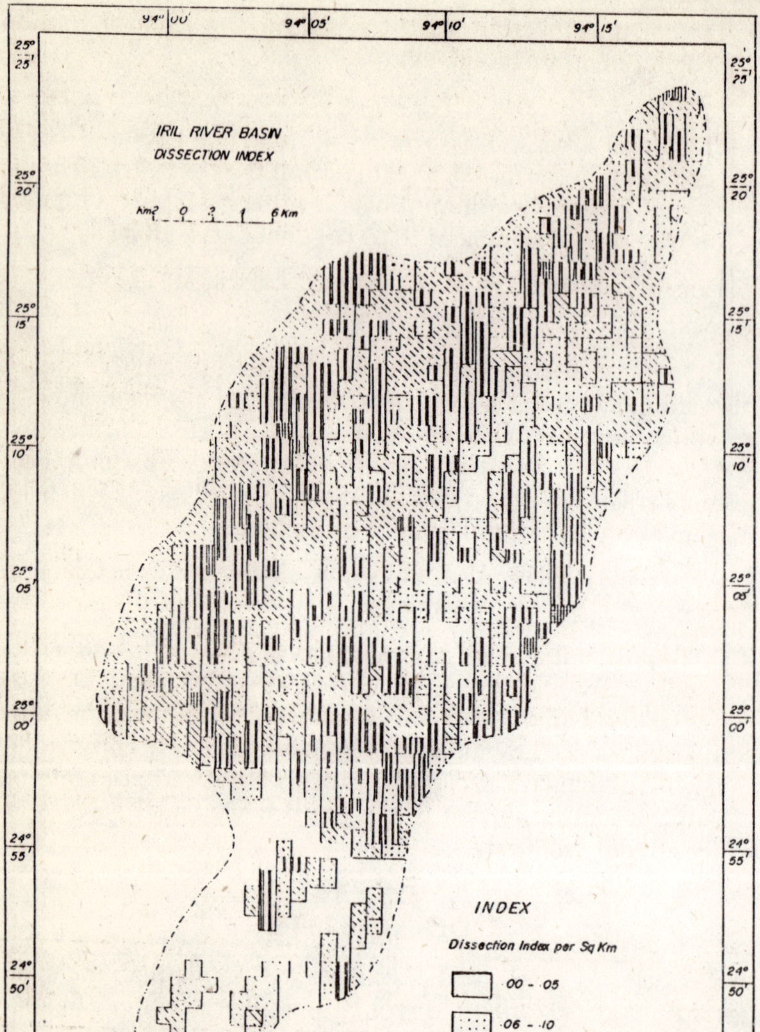
IRIL RIVER BASIN
DISSECTION INDEX
Km 2 0 2 4 6 Km
94° 00'
94° 05'
94° 10'
94° 15'
25° 25'
25° 20'
25° 15'
25° 10'
25° 05'
25° 00'
24° 55'
24° 50'
24° 45'
INDEX
Dissection Index per Sq Km
·00 - ·05
·06 - ·10
·11 - ·20
·21 - ·30
·31 - ·40
Above ·41

Fig. 5.4

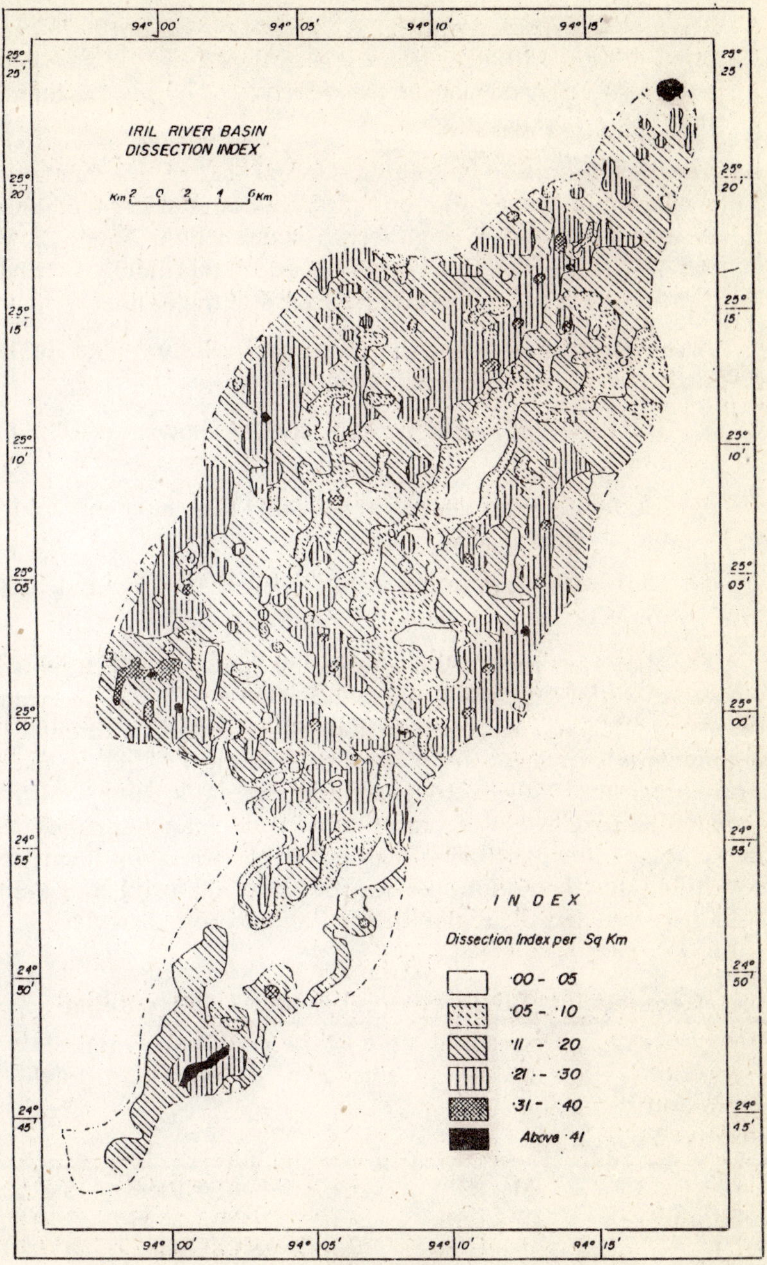
IRIL RIVER BASIN
DISSECTION INDEX
Km 2 0 2 4 6 Km
94° 00'
94° 05'
94° 10'
94° 15'
25° 25'
25° 20'
25° 15'
25° 10'
25° 05'
25° 00'
24° 55'
24° 50'
24° 45'
INDEX
Dissection Index per Sq Km
·00 - ·05
·05 - ·10
·11 - ·20
·21 - ·30
·31 - ·40
Above ·41

Fig. 5.4a

b. The medium dissection index is dominant in the northern and north eastern part of the basin. It is scattered all over the basin except the valley portion, *i.e.* the area covered by low dissection index extensively.

c. The high dissection index area is found scattered throughout the basin. However, the south eastern boundary of the basin is almost covered by high index of dissection, About 28.67 per cent of the total basin area falls under this index covering almost all the relief regions and scarps of the basin.

A comparison of dissection index and relative relief maps distinctly defines the following facts :

a. The region of low index of dissection shows low relative relief.

b. Moderate relief regions at the same time have moderate index of dissection.

c. High relief regions are also the regions of high index of dissection.

The above facts are well supported by Table 5.5, which shows a comparison of dissection index and relative relief data. From the above table, we can conclude that the area of high altitude, *i.e.* highlands with scarps, are well dissected and eroded by their drainage lines. The drainages have cut back their valleys broadening their ways and forming narrow elongated ridges in many areas, thus expressing high index of dissection. It can be concluded that the regions between 300m to 500m and above are well dissected and the region below 100m are less dissected.

TABLE 5.5

Comparison of Dissection Index and Relative Relief

Categories of Dissection Index and Relative Relief	Dissection Index	p.c. of the total area	Relative Relief	p.c. of the total area
Low	.00 - .05	16.67	below 100m	15.41
Medium	.05 - .20	54.66	100m - 300m	61.51
High	.20 - above .40	28.67	300m - above 500m	23.08

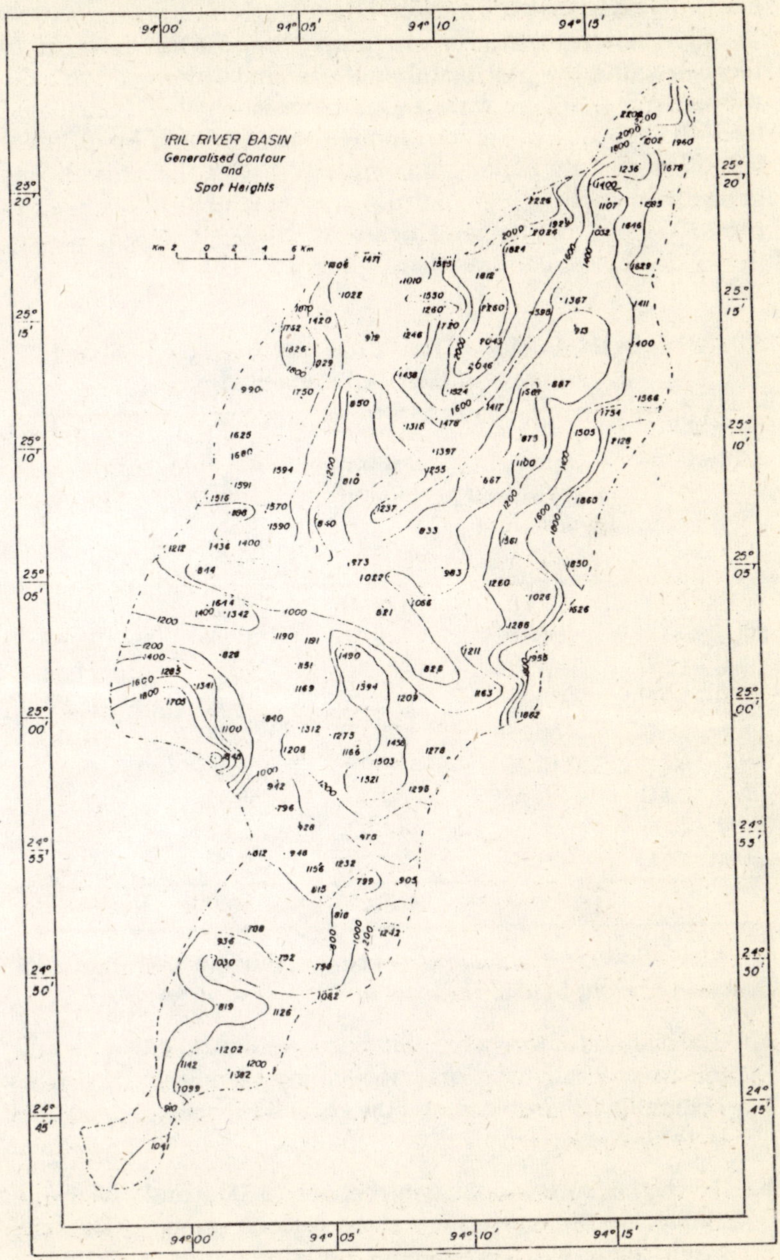
IRIL RIVER BASIN
Generalised Contour and Spot Heights
Km 2 0 2 4 6 Km

Fig. 5.5

5.5 GENERALISED CONTOUR

The generalised contour map (Fig. 5.5) is helpful in reconstructing the past features of the land surfaces. Table 5.6 shows the former land surface area, reconstructed and calculated from the generalised contour in the basin, enclosed by different height ranges. By making a comparison with the areas in the generalised contour map and the present land surfaces, we can identify some of the landforms in different height ranges destructed by erosion, weathering, etc.

TABLE 5.6
Areas in the Enclosed Two Contours from Generalised Contours of Past and Present Areas

Height ranges (m)	Area in past Generalised (Contour) (sq km)	Area at present (sq km)	% of the total area (past)	% of the total area (present)
Below 800	70.00	178.24	5.44	13.87
800 - 1000	230.96	342.58	17.98	26.66
1000 - 1200	269.20	271.75	20.95	21.13
1200 - 1400	252.25	202.48	19.63	15.76
1400 - 1600	186.30	129.80	14.50	10.10
1600 - 1800	166.20	79.27	12.93	6.17
1800 - 2000	81.09	60.36	6.31	4.69
2000 - 2200	24.00	18.57	1.87	1.45
above 2200	5.00	2.25	0.39	0.17
	1285.00	1285.00	100.00	100.00

The following conclusions can be derived after comparing the past and present land surface areas :

a. The higher surfaces in the past have been destructed by various agents of weathering, erosion etc. and therefore, these areas are gradually shrinking and the eroded substances deposited in the lower region.

b. The higher surface area, lying between 2000m and 2200m and above, which shows high steep regions today, previously might have had less steeper faces, and due to different erosional agents got their toe areas excavated and deposited in the next lower parts.

c. The area in between 800m and 1000m seems to have gained a lot as compared to the previous area. Due to continued destruction by the tributaries of Iril river and other agents the higher surfaces have been eroded and brought down and deposited in the lower relief region, and thus their slopes have also gradually decreased. This section shows the highest gain of area than the other height regions.

d. The lowest surface area, *i.e.* below 800m also received much extension, more than double of the past.

e. In comparison with the high relief, the lower surfaces have faced least destruction and received extension in area from the erosion of higher dissected and eroded parts.

5.6 STREAM LINE SURFACES

By interpolation of stream line surfaces, an indication of the amount of denudation yet to be achieved can be obtained. The stream line surface is that on which the available relief is zero. It is defined by contours passed through the intersections of actual contours with principal streams (Dury, 1952). In the stream line surface map (Fig. 5.6) the dissection that has been carved along the scarps comes out distinctly giving an idea of even future dissection. The stream line surface map of the Iril river basin distinctly shows the following facts :

a. The lower elevation areas, *i.e.* the middle and southern portions of the basin area where widespread *jhuming* is observed are heavily subjected to denudation with distinctly marked erosional surface.

b. The eastern and southern highland parts of the basin are prone to denudation.

c. The highest stream line surfaces are found in the north eastern part of the Iril basin.

d. The medium high and lowest stream line surfaces cover extensively the areas in the middle and southern parts of the basin.

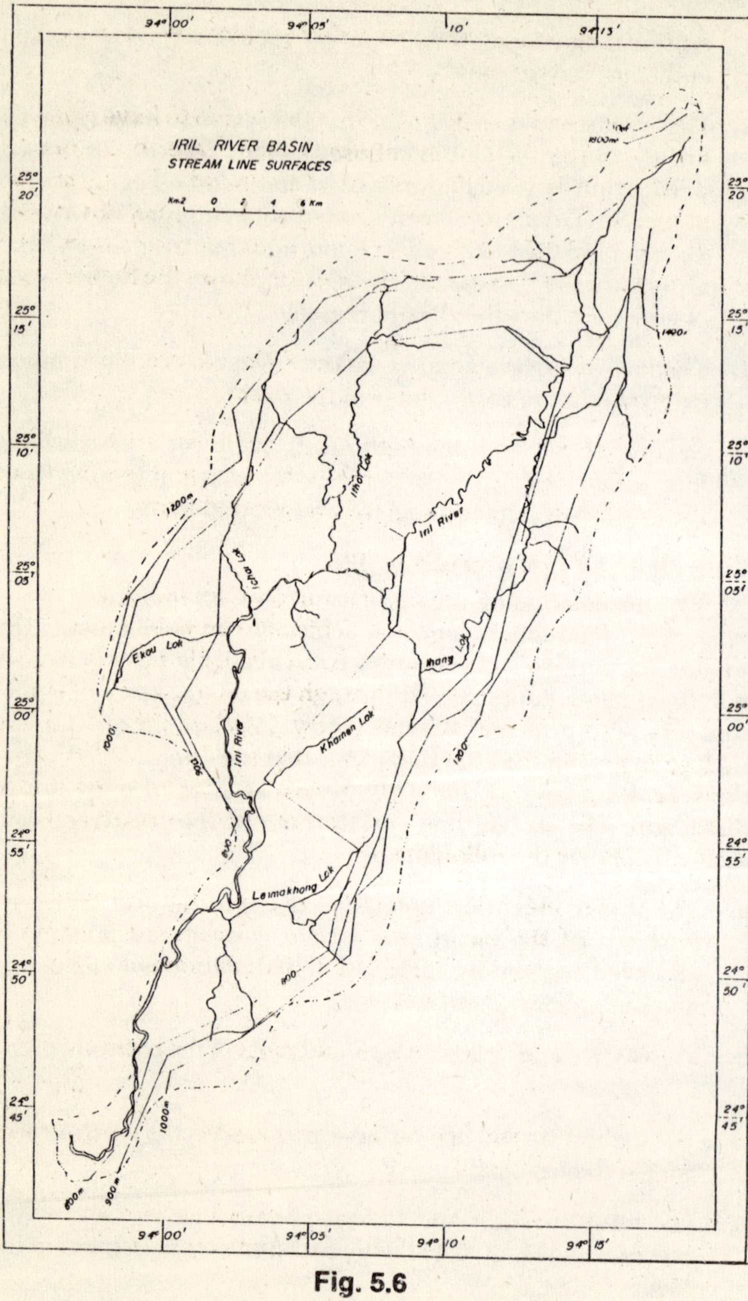
IRIL RIVER BASIN
STREAM LINE SURFACES
Km 2 0 2 4 6 Km
94° 00'
94° 05'
10'
94° 15'
25° 20'
25° 15'
25° 10'
25° 05'
25° 00'
24° 55'
24° 50'
24° 45'
94° 10'
1800m
1400m
1200m
1000m
900m
800m
Ithai Lok
Iril River
Ichai Lok
Ekou Lok
Ihang Lok
Khamen Lok
Leimakhong Lok

Fig. 5.6

5.7 AVERAGE SLOPE

The average slope determination method suggested by Raisz and Henry (1937) has been applied to this basin (Fig.5.7). Altogether seven average slope regions have been delineated. All categories show the amount of rise and fall of the surfaces in metre per km and the gradient of slope. Table 5.7 clearly indicates the six categories of slope with their gradient and broad classification.

Table 5.7
Slope Categories

Slope categories metre/km	Slope gradient	Slope in degree	Major slope categories
0 - 100	0 - 1/9.5	0 - 6	Gentle slope
100 - 200	1/9.5 - 1/5.2	6 - 11	Moderate
200 - 300	1/5.2 - 1/3.3	11 - 17	slope
300 - 400	1/3.3 - 1/2.6	17 - 22	Steep
400 - 500	1/2.6 - 1/2.2	22 - 26	slope
500 and above	1/2.2 and above	26 and above	Very steep slope

Fig. 5.7 delineates a broad classification of slope areas. The steep slope or the areas having gradient 1/2.6 to 1/2.2 or 22^0-26^0 slope are mostly scattered in the south western, north western and eastern parts of the basin area. The moderate or medium slope areas, with the slope between 200 and 300 m per km or 11^0 - 17^0 gradient, occupies most of the north western and some middle southern parts of the basin. The lowest value of slope, *i.e.* 0 - 100 m per km is found in the southern plain portion of the basin. This slope region forms a narrow elongated belt within the central portion of the basin which is extended upto the mouth of the Iril river.

The slopes, thus found out, are classified into 12 classes and regrouped into five slope categories such as gentle slope, moderate slope, moderately steep slope, steep slope and very steep slope.

Table 5.8 shows that the Iril basin is manifested with various degrees of slope. About half of the basin area (49.08 per cent) has

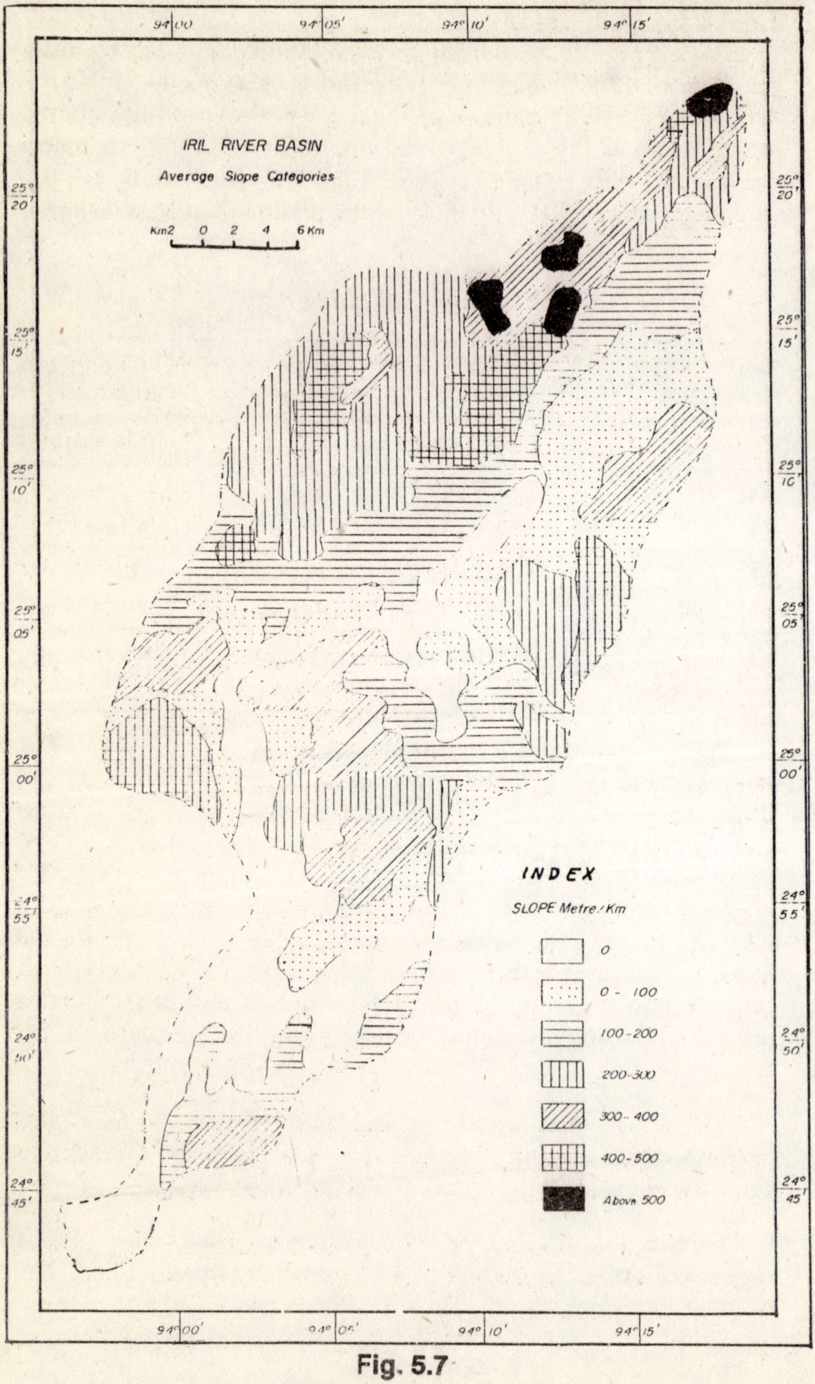
IRIL RIVER BASIN
Average Slope Categories
Km2 0 2 4 6 Km
INDEX
SLOPE Metre/Km
0
0 - 100
100-200
200-300
300-400
400-500
Above 500
94°00'
94°05'
94°10'
94°15'
25° 20'
25° 15'
25° 10'
25° 05'
25° 00'
24° 55'
24° 50'
24° 45'

Fig. 5.7

steep slope, while the very steep slope covers about 6.33 per cent of the basin. These slope areas are followed by moderately steep slope, moderate slope and gentle slope categories. In comparison with dissection index, it is seen that the higher area, having high index of dissection, shows high degree of slope. Most of the areas are under the steep slope and moderately steep slope regions, thus expressing high intensity of dissection.

Areas of gentle slope are the least as shown by the average slope map. Isopleth map (Fig. 5.8) shows the trend of area having equal degree of slope. The source area of the Iril river has very steep slope, while the lower or the mouth part of the river has gentle slope with 4^0 - 6^0 gradient. The area under very steep slope, ranging between 24^0 to 28^0, tends to lay along with more dissected relief regions.

Table 5.8
Average Slope

Average slope in degree	Frequency (sq. km.)	% of the total basin area	Cumulative Frequency	Major slope categories
4 - 6	40.71	3.17	3.17	Gentle slope 6.02%
6 - 8	36.58	2.85	6.02	
8 - 10	47.48	3.69	9.71	Moderate
10 - 12	66.25	5.15	14.86	slope 8.84%
12 - 14	27.12	2.12	16.98	
14 - 16	68.65	5.35	22.33	Moderately
16 - 18	65.41	5.09	27.42	steep slope 29.73%
18 - 20	220.70	17.17	44.59	
20 - 22	322.50	25.09	69.68	Steep slope
22 - 24	308.30	23.99	93.67	49.08%
24 - 26	63.10	4.91	98.58	Very steep
26 - 28	18.20	1.42	100.00	slope 6.33%

A study of average slope and slope categories reveals the slope

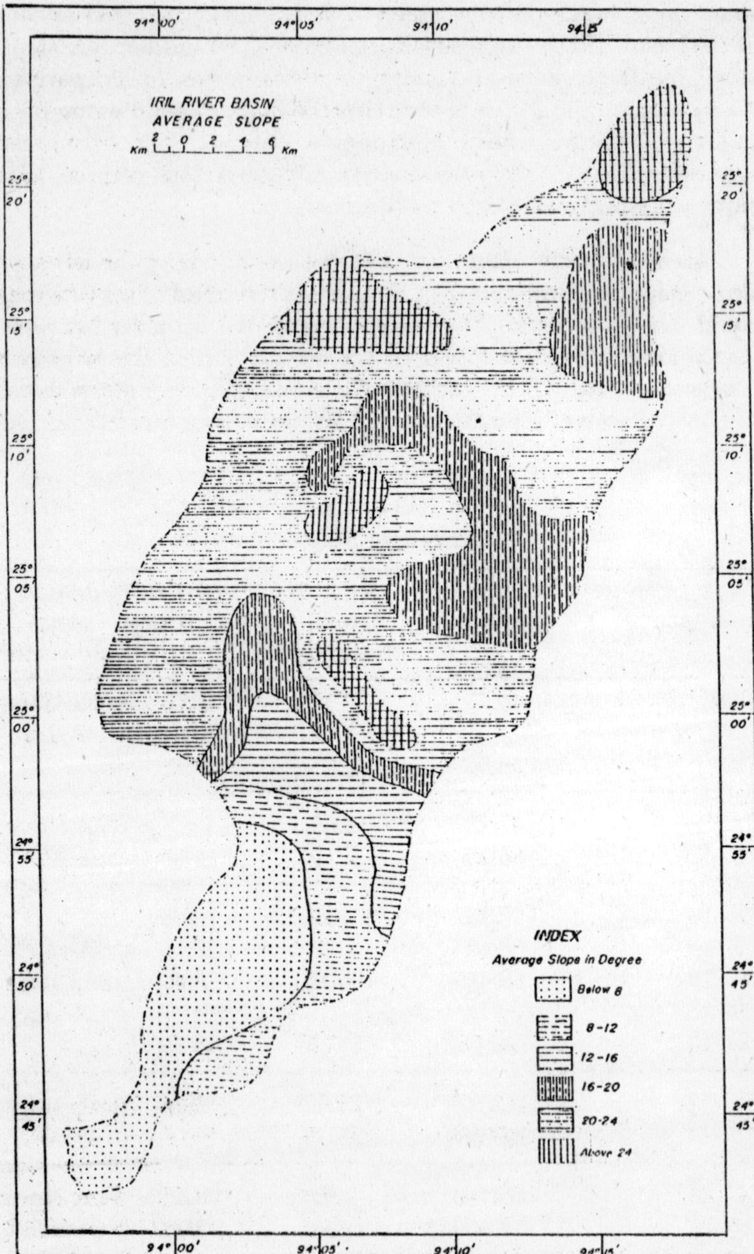
IRIL RIVER BASIN
AVERAGE SLOPE
Km 2 0 2 4 6 Km
94° 00'
94° 05'
94° 10'
94° 15'
25° 20'
25° 15'
25° 10'
25° 05'
25° 00'
24° 55'
24° 50'
24° 45'
INDEX
Average Slope in Degree
Below 8
8-12
12-16
16-20
20-24
Above 24

Fig. 5.8

relationship, which represents the character of the area with reference to its slope region. Slope Ogive (Fig. 5.9), showing the distribution of slope, indicates a gradual rise of slope upto the level of 18^0 and its sudden inclination covering a large area upto 24^0. Its further rise is less gradual expressing lesser area under very steep slope category.

5.8 SLOPE REGIONS

From the above analysis of slope and relief, as expressed by Eric (1967), frequency of slope magnitude may be ascertained and the aerial distribution of slope values portrayed. Most of the lower slope regions are found along the central, upper southern and mouth parts of the Iril river. The north eastern and south western areas show composite nature of terrain, which varies in slope from medium to high. However, medium slope region is much extensive than the high slope region. A broad belt of medium slope runs in the more or less central part from the northern to south - western part dotted by low relief regions in some parts. The frequency values of lower relief regions are concentrated along the broad valleys attaining mature stage. Therefore, the steep slope and the moderate slope regions have developed in the well dissected scarp regions and over hilly terrain under the impact of fluvial system. By examining the aerial distribution of slope values we can identify three types of slope regions in the Iril basin area.

5.8.1 GENTLE SLOPE AREA

The areas of gentle slope, with a gradient of 8^0 and below, occupy around 6.02 per cent of the total area of the Iril basin. This gentle slope area, which coincides with the low relief region, low to moderate texture of dissection and low drainage density, extends in the lower southern portion upto the mouth of the Iril river. It is surrounded by moderate slope areas.

5.8.2 MODERATE TO MODERATELY STEEP SLOPE AREA

The slope 8^0 - 20^0 comes under the category of moderate to moderately steep slope area, and covers most of the southern half of the basin. Big patches of moderately steep slope area are also found in the lower northern and upper north western parts of the basin. This category of slope covers an area of 38.57 per cent of the basin. It is the second largest slope region coinciding with medium to highly dissected landscape.

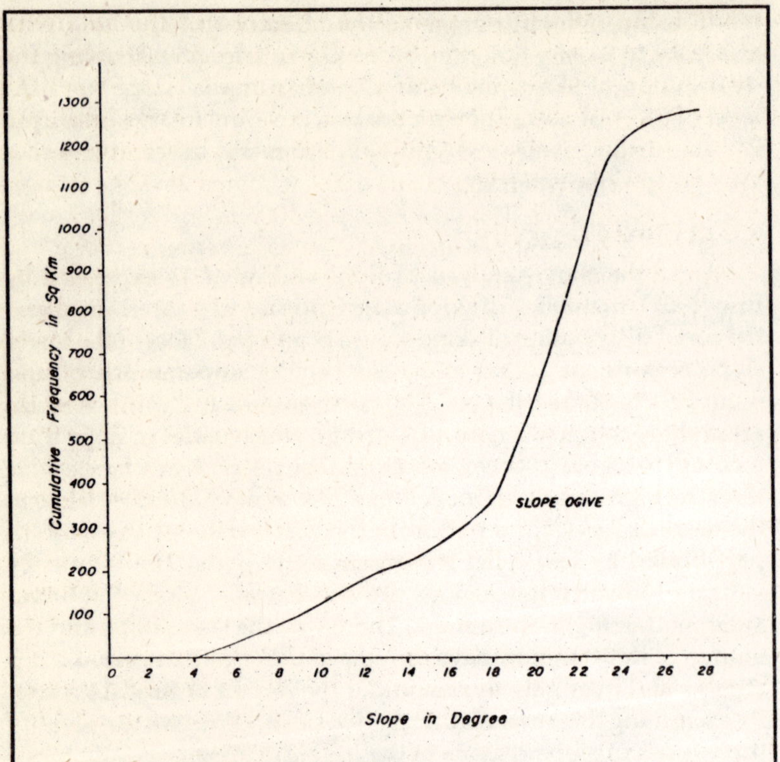

Fig. 5.9 Relationship of Area and Slope

5.8.3 STEEP TO VERY STEEP SLOPE AREA

The steep slope region covers around 55.41 per cent of the total basin area. All the hilly and highly dissected areas are covered by these categories of slope ranging from 20⁰ to 28⁰. The source area of the Iril river, the extreme northern part of the basin, comprise of very steep slope. Most of the western and north - western parts are also covered under this category of steep slope. This highly dissected area has a high drainage density.

Different magnitudes of slope have been identified in the basin. The elevated valleys have gentle slope while the displaced surfaces have steep slopes. Thus, the Iril river basin has a close correlation between the relief, drainage texture, dissection and slope, and as such is the combined result of the earth movement, fluvial system and various agents of erosion and weathering.

5.9 ALTIMETRIC FREQUENCY HISTOGRAM

The frequency of occurrence of different elevation groups in a region can be clearly shown in an altimetric frequency histogram. The altimetric frequency histogram of Iril river basin (Fig. 5.10) presents a clear picture of the relative position and percentage of different land surfaces of Iril basin. From this diagram it is seen that the area between 800 m and 1000 m covers 26.82 per cent of the total basin area. The area between 1000 m and 1200 m cover about 21.12 per cent of the basin. The area between 800 m and 1000 m and below covers all the plain portion of the Iril basin. The region having the contour interval of 2000 m to 2200 m and above are of limited extent generally forming hill slopes and steep scarps covering a small area of the basin, *i.e.* 1.41 per cent and 1.17 per cent respectively. These portions are almost confined to the northern and north eastern parts of the basin area.

5.10 HYPSOMETRIC OR HYPSOGRAPHIC CURVE

The study of hypsometric or hypsographic curve can reveal the stage reached by a drainage basin in the present cycle of erosion and the percentage of mass removed from the basin by different erosional processes. The area under the hypsometric curve is known as the hypsometric integral which is a measure of stage "because it expresses as a percentage of the mass of the drainage basin remaining above a basal plane of reference" (Schumm, 1956).

All the main drainages of Iril river are at the stage of maturity except Wari Lok showing 72.05 per cent of hypsometric integral value, thus showing a lot of land surface to be eroded in the near future. The drainages Ithoi, Ihang and Leimakhong are at the marginal zone of maturity and monadnock phase showing 49.33 per cent, 48.06 per cent and 47.76 per cent respectively. The drainages Ekou and Ichai are also at their maturity stages although they have somewhat higher hypsometric integral values *i.e.* 51.54 per cent and 57.62 per cent. The hypsometric integral value for the Iril river is 37.77 per cent, although its value is slightly over the rough limit of 35 per cent set by Strahler (1952), yet other physical factors confirm its senility and showing the monadnock phase.

Hypsometric curve is also used to indicate the proportion of the area of the surface at various elevations above or below a given datum. It indicates the form of quality of erosional surface and depicts more or less the exact stage of development of the basin. This curve "relates horizonal cross sectional area of a drainage basin to relative elevation above basin mouth" (Strahler, 1952). From this curve it is seen that the area between 2000 m to 2200 m height and above are very limited and may be called as highland zone. The area between 1200 m to 2000 m height covers the main basin of the Iril river with scattered sections of steep slopes. The land between 1000 m to 1200 m has medium high gradient with considerably high erosional surfaces, while the area with elevation of 900 m and below form some what flat plain land and receive masses removed by erosion (Figs. 5.11 to 5.11c). Table 5.9 and 5.10 show the areas of the different drainage basins enclosed by the consecutive contours in Iril river basin and their areas in per cent with cumulative percentage.

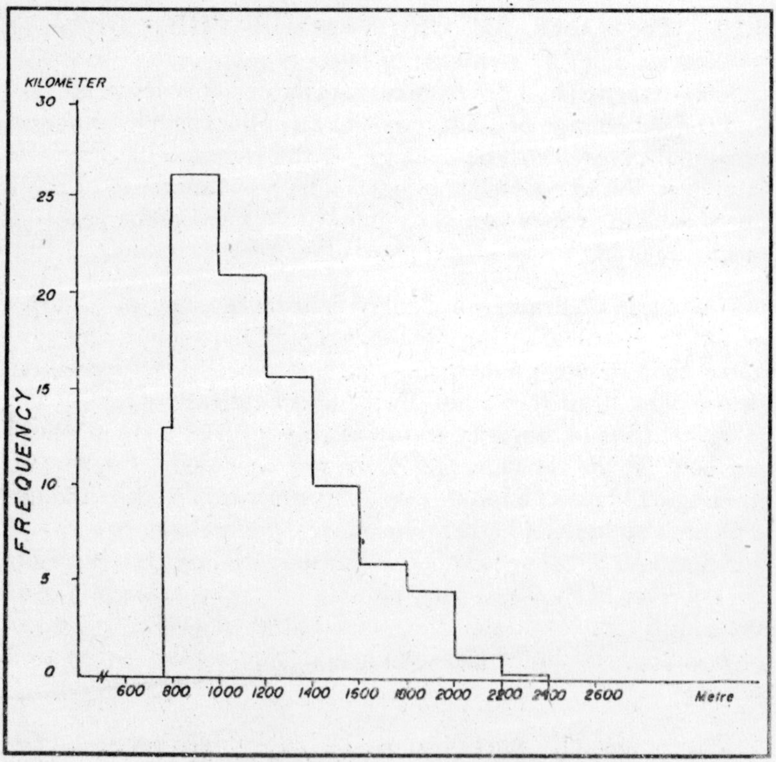

Fig. 5.10 Altimetric Frequency Histogram

Table 5.9

Area Enclosed by the Consecutive Contours in Iril River Basin

Contour Group (m)	Iril	Ithoi	Ihang	Leima-khong	Ekou	Ichai	Wari
Above 2200	2.25						0.19
2000 - 2200	18.57	0.60	1.93				7.73
1800 - 2000	60.36	9.45	10.20				8.38
1600 - 1800	79.27	17.61	16.38		1.02	3.25	7.25
1400 - 1600	129.80	33.60	22.75		7.74	9.86	5.81
1200 - 1400	202.48	58.33	38.59	4.89	11.99	12.50	3.78
1000 - 1200	271.45	71.94	53.41	14.90	15.56	16.84	1.71
800 - 1000	342.58	43.14	52.40	52.12	19.65	14.71	0.12
below 800	178.24	2.83	00.59	40.12	0.18	1.86	
	1285.00	237.50	196.25	112.03	56.14	59.02	34.97

5.11 AREA HEIGHT CURVE

The area height curve shows the actual area of land between two adjacent contours, depending on the selected contour interval (Monkhouse and Wilkinson, 1980). In this curve for the Iril river basin (Fig.5.12), the percentages of the areas calculated have been plotted against the heights which are in absolute values. The land percentages above 2000 m elevation and in between 1600 m and 1900 m show a sharp diminishing trend, thus indicating active erosional activity with deep dissection and severe erosion.

5.12 THALWEG OR THE LONGITUDINAL PROFILE

The Thalweg or the longitudinal profile of a river portrays the relief, effects of uplift warping and subsidence and presents an outline of the valley surface from source to mouth. This is a "profile along the winding line of the valley floor" (Miller, 1953). Longitudinal profile is a graph of distance versus elevation which is function of load discharge, size of debris, width, depth and slope.

The profile of Ichai Lok shows three breaks of slope which can identify three broad geomorphic surfaces, two of which at the level of 1000 m and the other showing a stable position afterwards. The Wari Lok, as a short hilly stream, shows sudden drop from the height of 2200 m upto 1000 m within a short distance of 6.5 km and joins the main stream within a short distance.

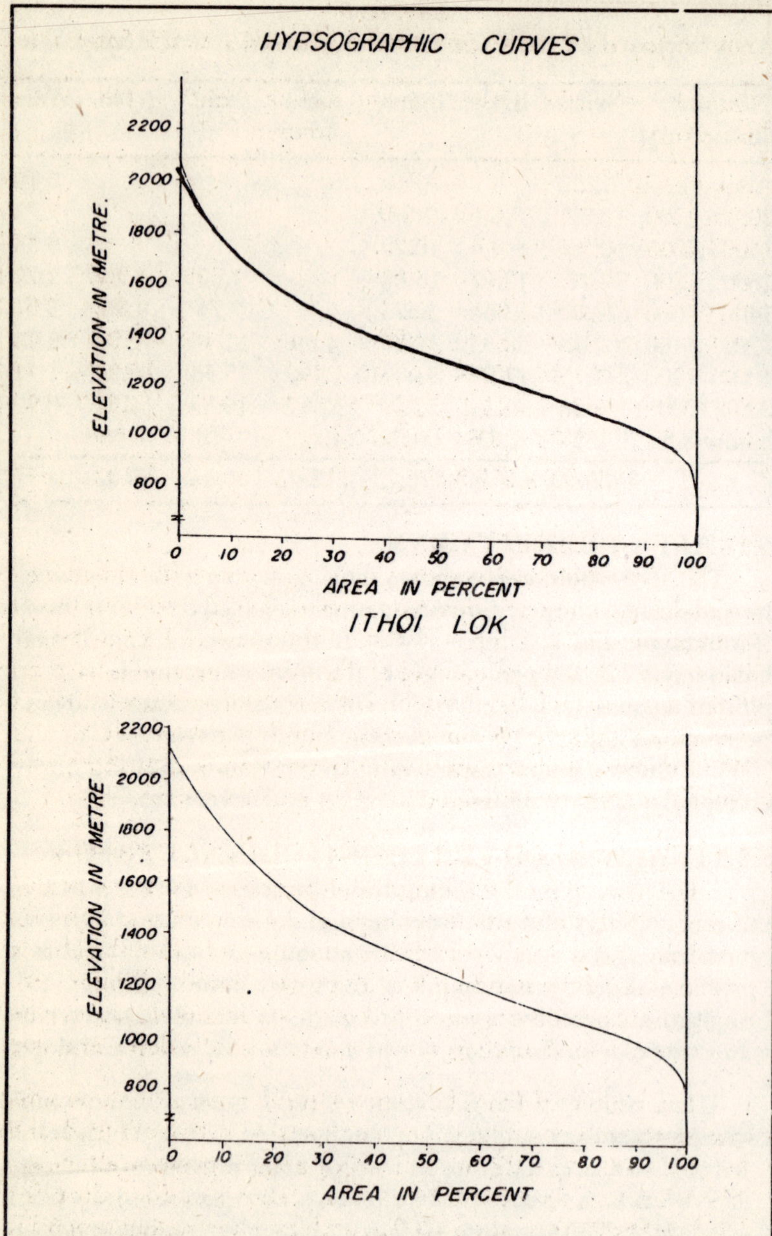

Fig. 5.11 IHANG LOK

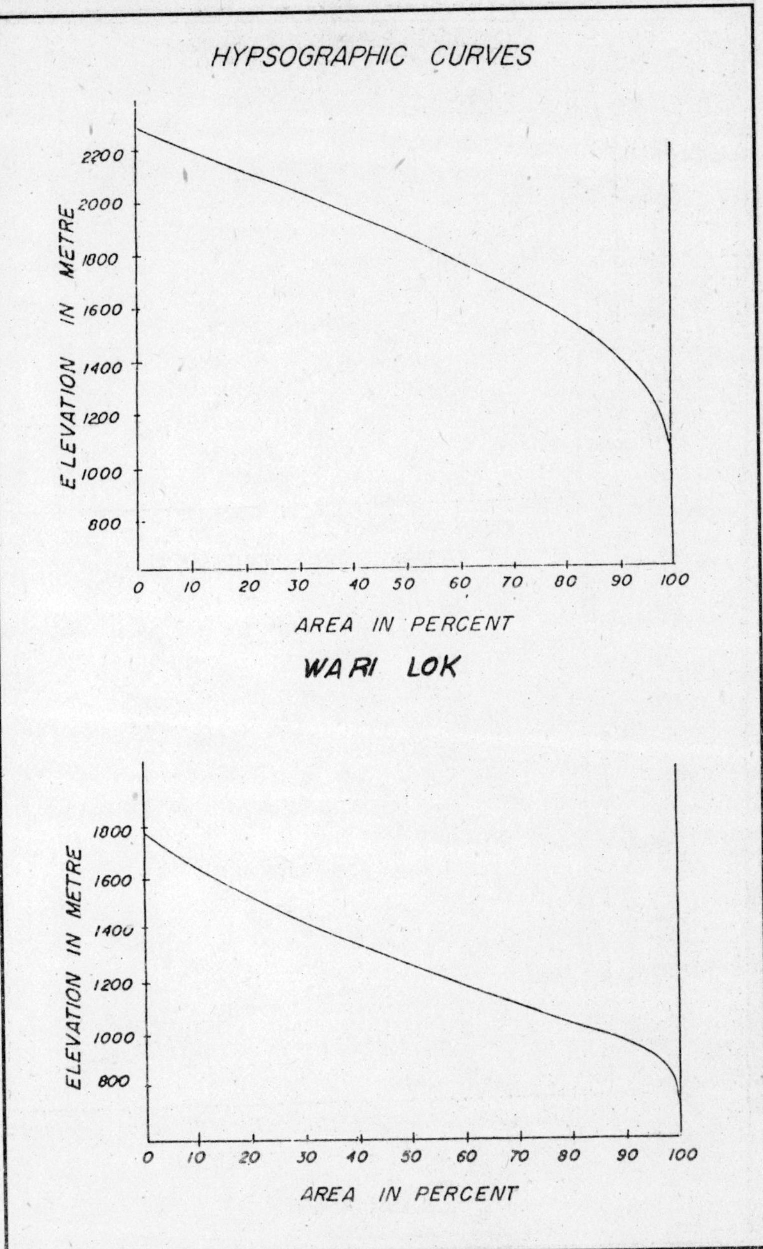

Fig. 5.11a ICHAI LOK

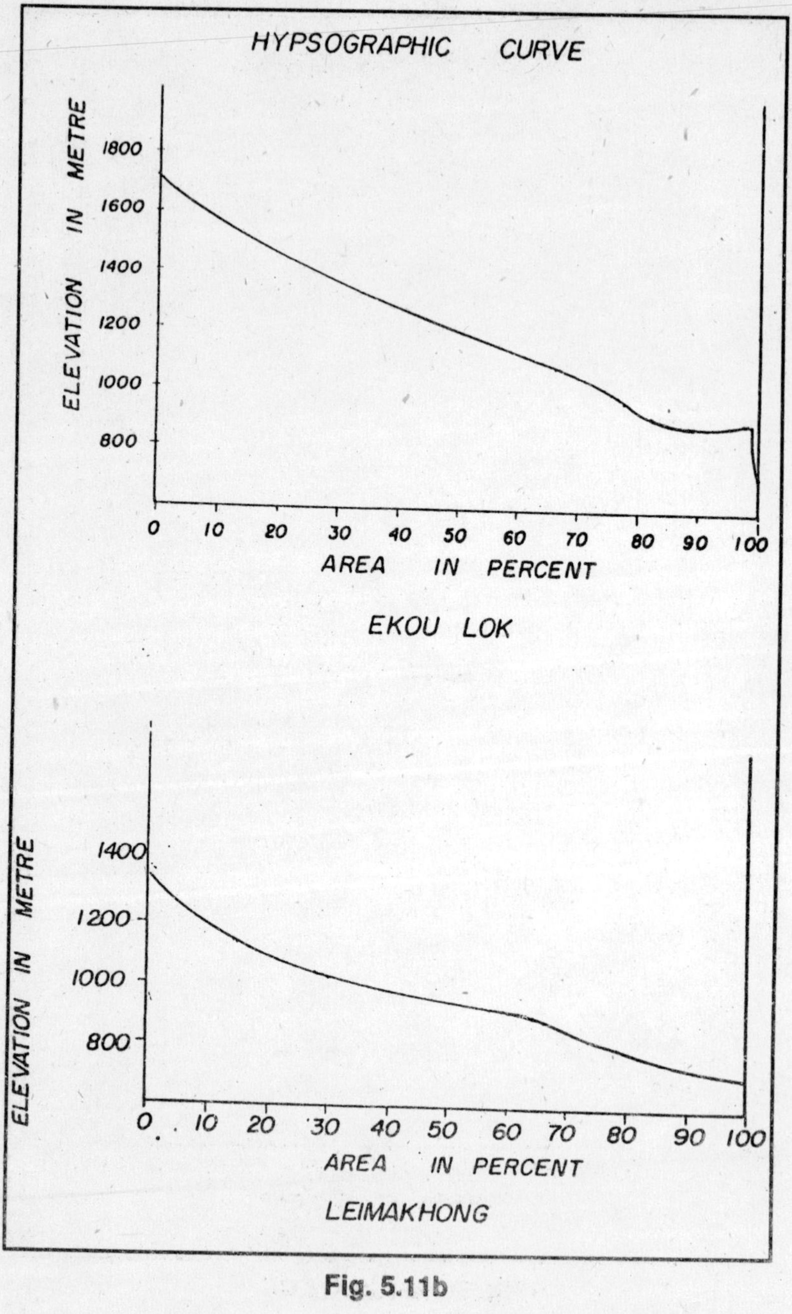
HYPSOGRAPHIC CURVE
ELEVATION IN METRE
1800
1600
1400
1200
1000
800
0 10 20 30 40 50 60 70 80 90 100
AREA IN PERCENT
EKOU LOK
ELEVATION IN METRE
1400
1200
1000
800
0 10 20 30 40 50 60 70 80 90 100
AREA IN PERCENT
LEIMAKHONG

Fig. 5.11b

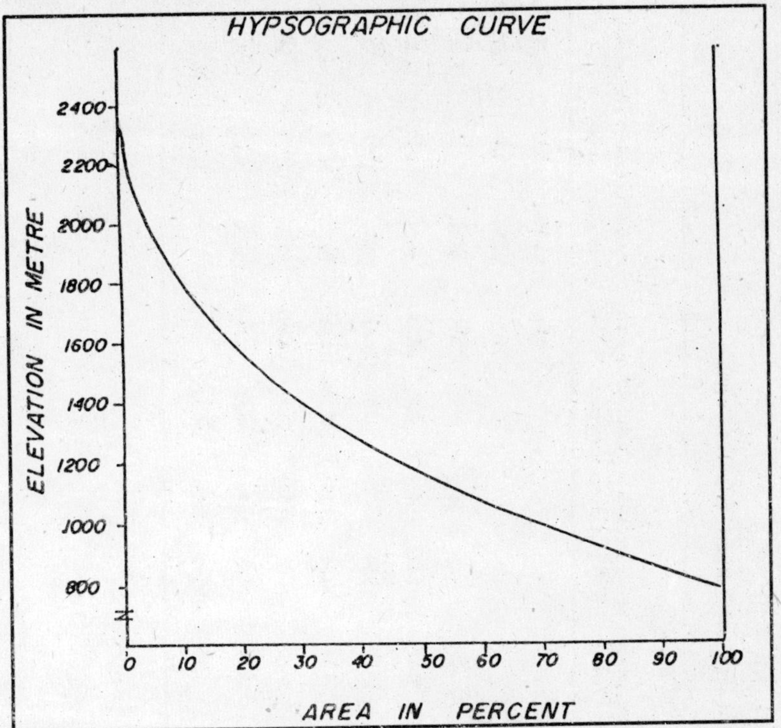

Fig. 5.11c IRIL RIVER

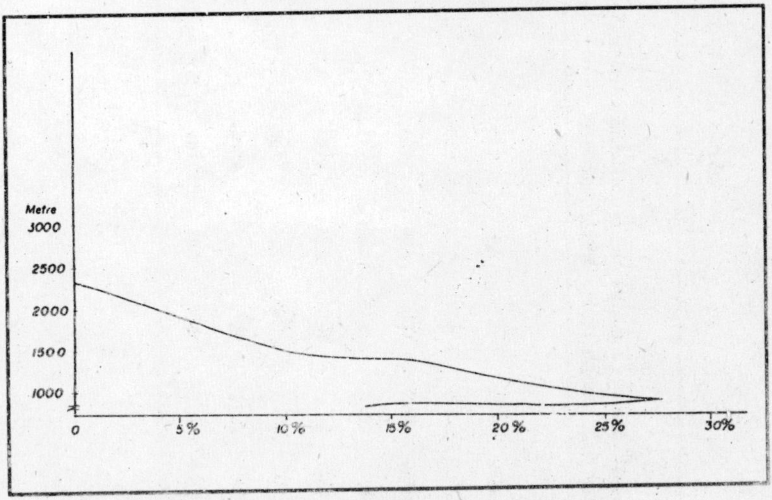

Fig. 5.12 Area Height curve of Iril River

Table 5.10
Area of Drainage in Iril Catchment (in per cent)

Contour	Ithoi		Ihang		Leimakhong		Ekou		Ichai		Wari		Iril	
Interval (in Metre)	Area in Percent	Cumulative percentage	Area in Percent	Cumulative Percentage	Area in Percent	Cumulative Percentage	Area in Percent	Cumulative Percentage	Area in Percent	Cumulative Percentage	Area in Percent	Cumulative Percentage	Area in Percent	Cumulative Percentage
above 2200											0.54	0.54	0.17	0.17
2000-2200	0.25	0.25	0.99	0.99							22.10	22.64	1.44	1.61
1800-2000	3.98	4.23	5.20	6.19							23.97	46.61	4.69	6.30
1600-1800	7.42	11.65	8.34	14.53			1.82	1.82	5.47	5.47	20.73	46.34	6.17	12.47
1400-1600	14.15	25.80	11.60	26.13			13.78	15.60	16.71	22.18	16.62	83.96	10.10	22.57
1200-1400	24.56	50.36	19.66	45.79	4.37	4.37	21.36	36.96	21.20	43.38	10.81	94.77	15.76	38.33
1000-1200	30.29	80.65	27.21	73.00	13.30	17.67	27.72	64.68	28.53	71.91	4.89	99.66	21.13	59.46
800-1000	18.16	98.81	26.70	97.70	46.52	64.19	35.00	99.68	24.94	96.85	0.34	100.00	26.67	86.28
below 800	1.19	100.00	0.30	100.00	35.81	100.00	0.32	100.00	3.15	100.00			13.87	100.00

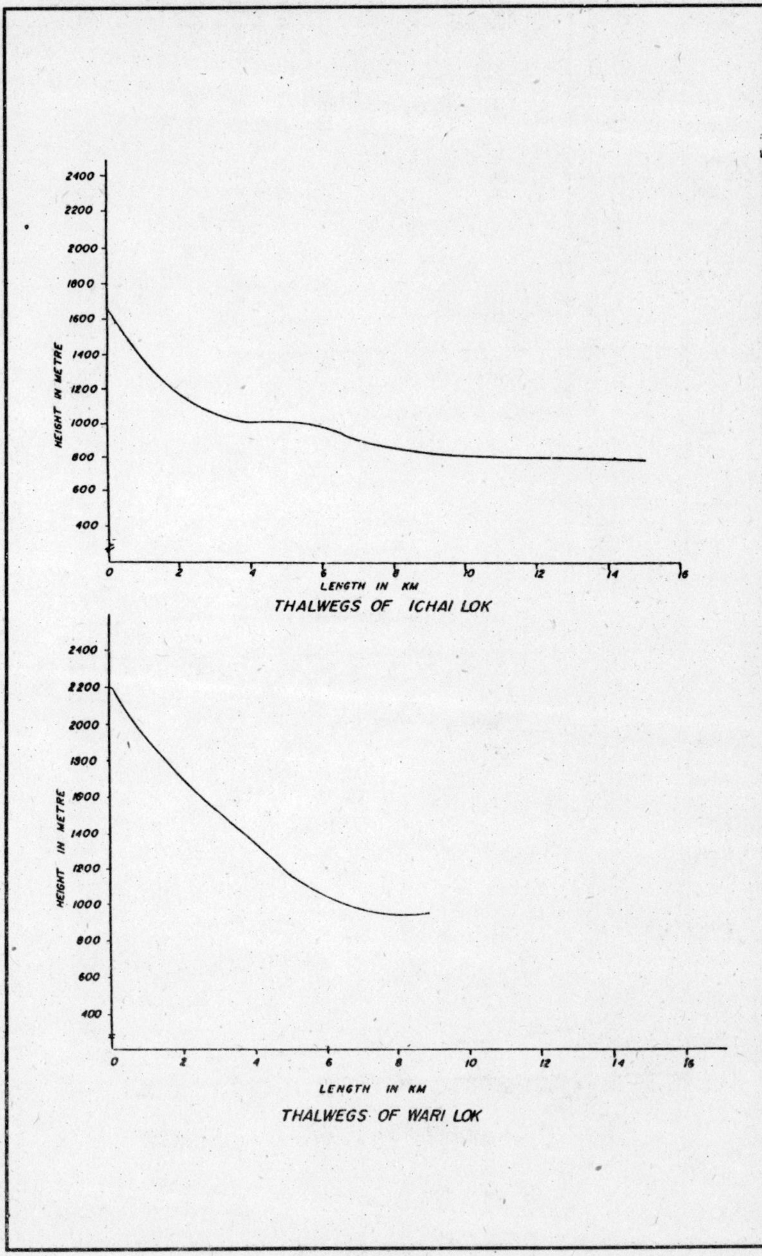
HEIGHT IN METRE
LENGTH IN KM
THALWEGS OF ICHAI LOK
HEIGHT IN METRE
LENGTH IN KM
THALWEGS OF WARI LOK

Fig. 5.13

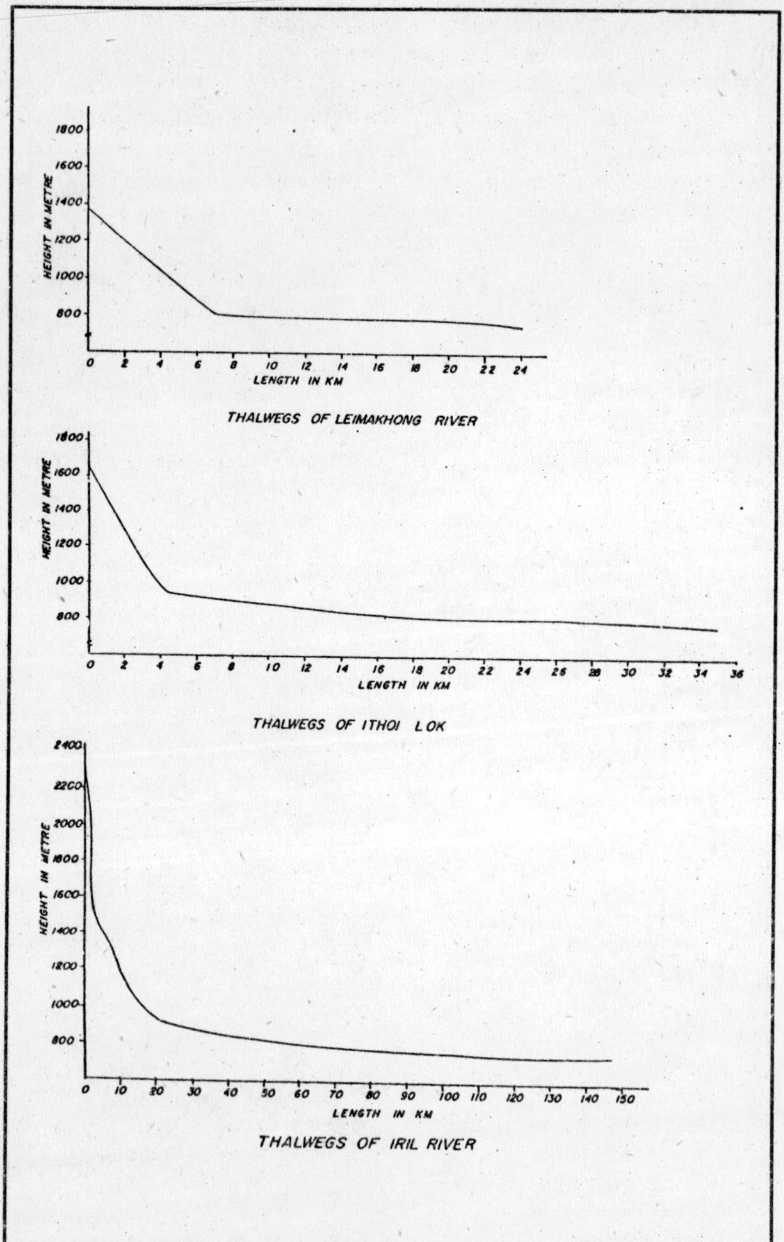
HEIGHT IN METRE
LENGTH IN KM
THALWEGS OF LEIMAKHONG RIVER
HEIGHT IN METRE
LENGTH IN KM
THALWEGS OF ITHOI LOK
HEIGHT IN METRE
LENGTH IN KM
THALWEGS OF IRIL RIVER

Fig. 5.13a

River Leimakhong shows sudden drop from its source and maintains its flow in the lower reaches showing two phases in its journey. River Ithoi also shows two phases of geomorphic surface. The Iril river also suddenly grades down its height within 10 to 20 kms and afterwards flowing almost at the plain attaining maturity stage and showing at least three phases of cycle in its journey. These are the higher gradient, the middle gradient and the lower gradient in the lower reaches (Figs. 5.13 - 5.13b). As most of the streams start from the high relief, they grade downward suddenly and attain the condition of a balance between erosion and deposition of a mature river.

5.13 CROSS PROFILE OF THE LANDSCAPE

The nature of a valley or a region, such as form, stage in the cycle of erosion, breaks of slope, terrace formation etc., can be identified by drawing a cross profile of the region. Here, Iril river basin have been divided into three parts and serial profiles have been drawn across the river from west to east and viewed from the south. The superimposed profile shows some morphological unity, such as the erosional surfaces and depicts the surfaces of different slopes. A projected profile will give a panoramic effect, with a distant skyline, thus showing only summit details. The composite profile can represent the surface of any area of relief which shows surface ruggedness and the general skyline of the landscape.

The landscape profile of the upper portion of the Iril basin (Fig. 5.14) reveals the following points :

a. Three erosional surfaces exist at 1000 m level, besides two erosional surfaces at the level of 900 m and one erosional surface at the level of 800 m.

b. The entire landscape with the height upto 2000 m and above can be called as highland areas.

c. The eastern parts of the basin have the higher elevation in comparison to the western parts.

d. Composite profile of the landscape shows that the area has been dissected by a number of streams. As such, this region appears to be a dissected highland.

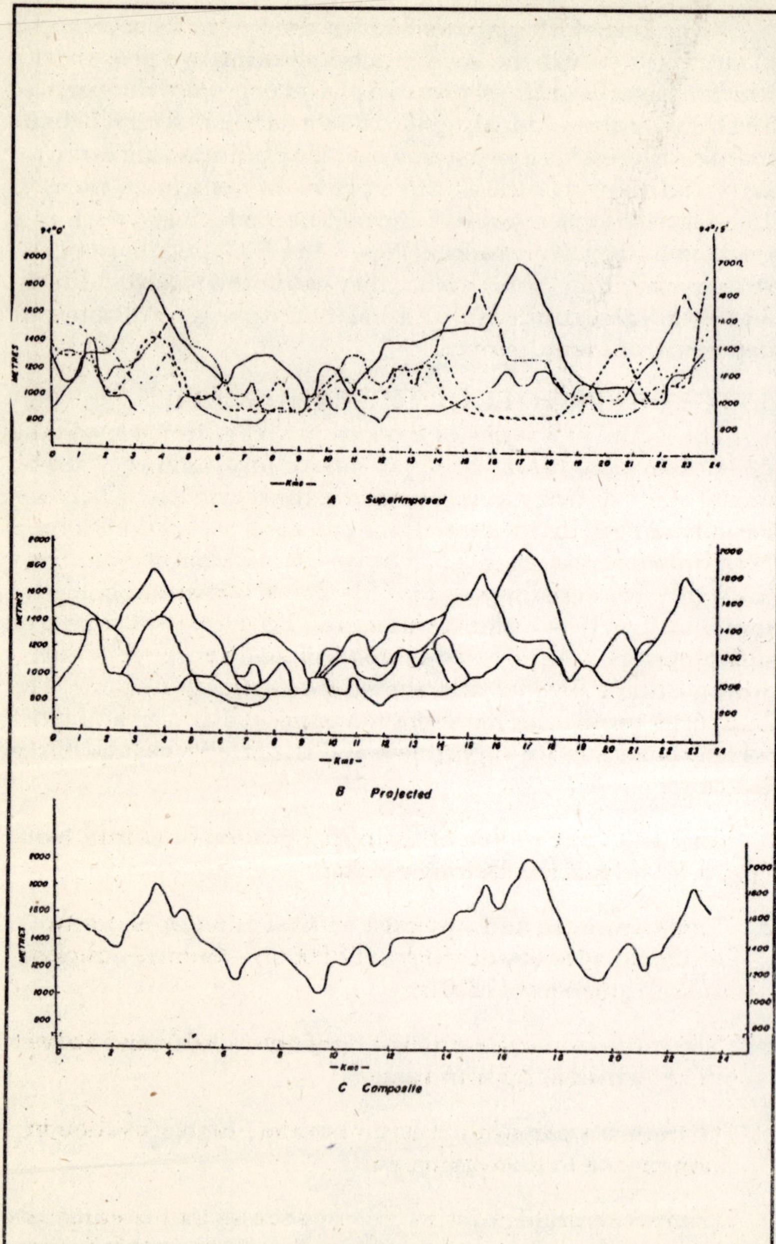

Fig. 5.14

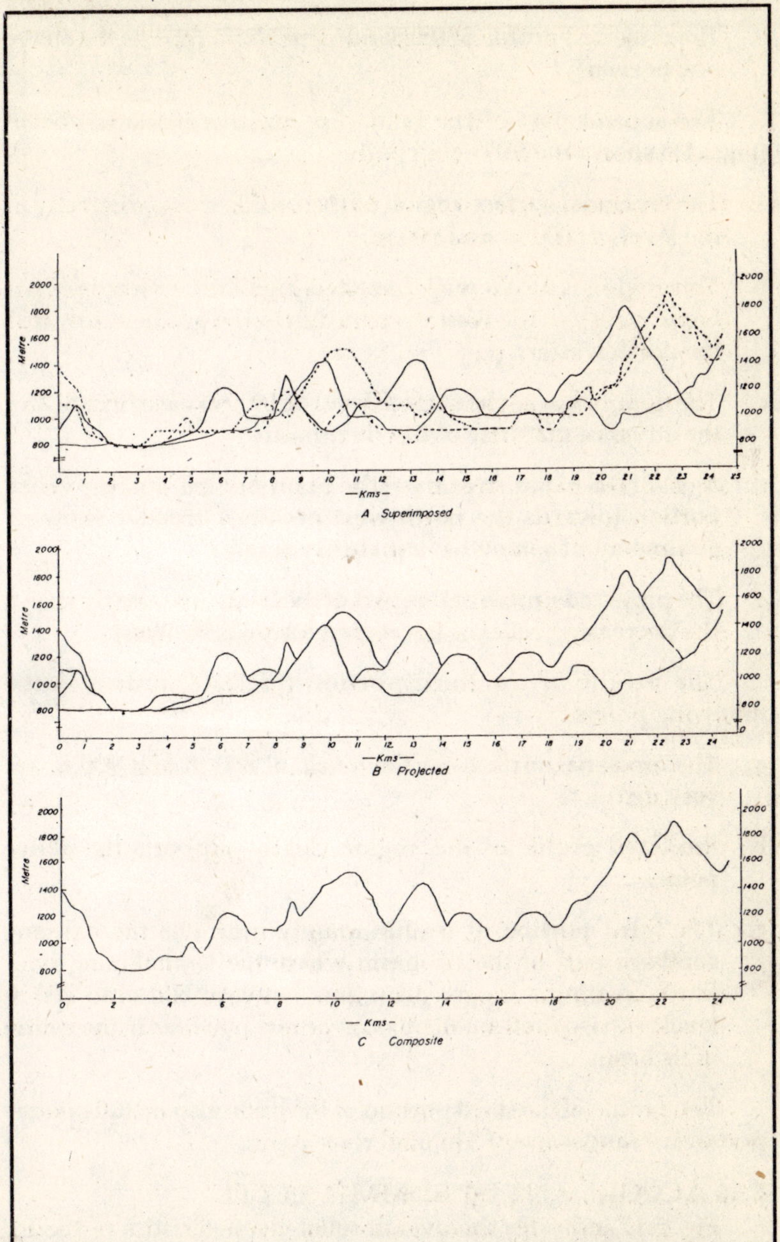

Fig. 5.15 Cross Profile of the landscape (Middle Part)

e. The projected profile of this landscape also supports the above points clearly.

The second part of the landscape profile of the Iril basin (Fig.5.15) shows the following points :

a. The erosional surface region of the profile exists distinctly at the levels of 800 m and 900 m.

b. This region is also a well dissected highland with somewhat broad valley in the western part due to the gradient towards the north western part.

c. The dominance of the surface level of 1000 m clearly indicates the advance maturity of the river basin.

d. Most of the major streams of the basin join Iril, and the valley portion towards the north west becomes broader showing attainment of somewhat a maturity stage.

e. The projected profile of this part of the landscape clearly shows the decreasing relief of the region towards the west.

The profile of the third portion (Fig.5.16) indicates the following points :

a. The erosional surfaces at the levels of 800 m and 900 m are very distinct.

b. Projected profile of this region clearly supports the above points.

c. The third portion of profile mainly indicates the extreme southern part of the Iril basin where the Leimakhong river flows creating a bigger plain area between 800 m to 900 m level with isolated small hillocks or residual hills in the centre of its basin.

The profile of the third portion of the basin also includes some part of the landscape of Thoubal river basin.

5.14 ACCORDANT OF SUMMIT LEVEL

Fig. 5.17 indicates the overall relief characteristics of the Iril basin. By combination of the projected profile of the 3 sets, some similar levels of land surface can be easily identified. This may be

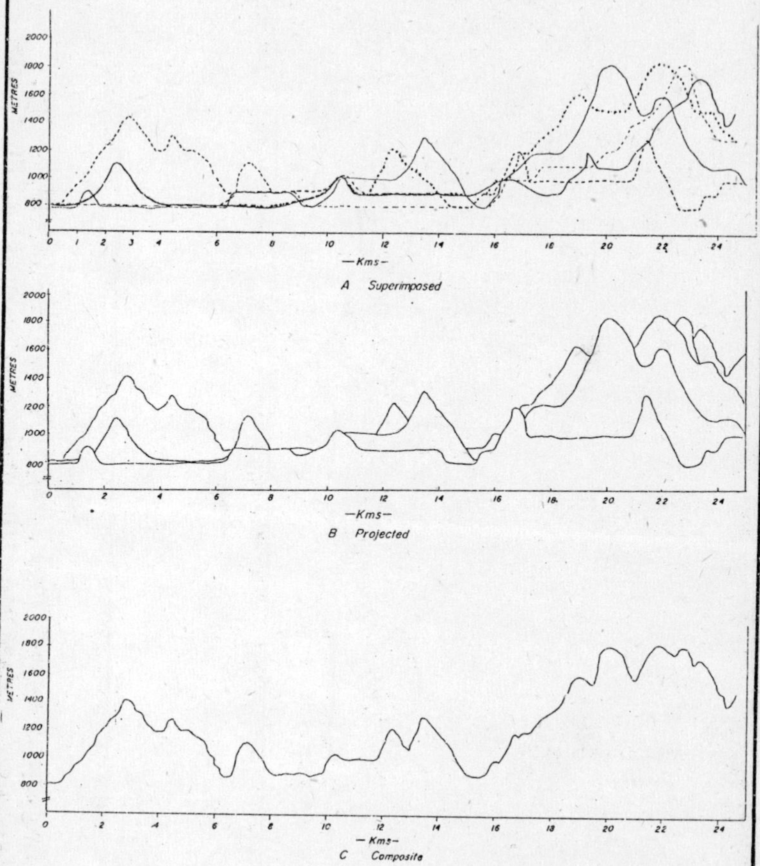
METRES
2000
1800
1600
1400
1200
1000
800
0 1 2 3 4 6 8 10 12 14 16 18 20 22 24
—Kms—
A Superimposed
METRES
2000
1800
1600
1400
1200
1000
800
0 2 4 6 8 10 12 14 16 18 20 22 24
—Kms—
B Projected
METRES
2000
1800
1600
1400
1200
1000
800
0 2 4 6 8 10 12 14 16 18 20 22 24
— Kms—
C Composite

Fig. 5.16

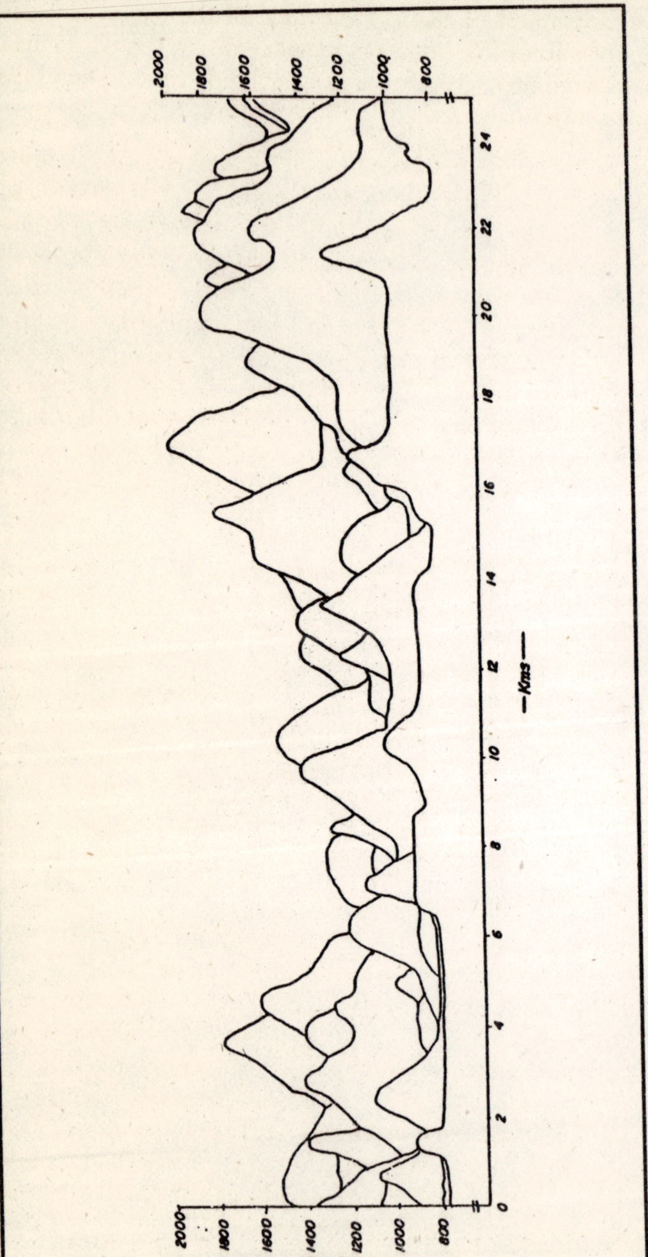

Fig. 5.17
Projected Profile of the Iril basin / Accordant of Summit Level
(Prepared by combination of the Projected Profile of the Three Parts)

due to successive intermittent uplifts during the Tertiary period with the movement of the land surface of the basin. However, remnants of successive attempts to wear down the land surface might have been preserved in the present landscape as Dury (1969) opines on similar processes.

Here at the level of 1800 m, one accordant of summit level is well recognised. In the same manner at the level of 1600 m, 1400 m, 1280 m, 1200 m and 1000 m , other five accordants of summit level are also recognised. The erosional surface at the level of 1000 m and 900 m are very distinctly seen, which represent general relief towards the NW - SE direction. The undulating hills in the profile itself shows that the maximum part of the basin is dissected. Residual hills are distinctly seen in the profile along the Iril river. Besides these, there is another surface, the lowest of this region at the south western part at 800 m, showing the depositional surface with a very limited occurrence.

5.15 TRAVERSE VALLEY PROFILE

By drawing cross profile along the river beds, the nature and the form of the valley, the stage in the cycle of erosion, the breaks in slope either due to incomplete grading of slopes owing to differing degree of resistence or rejuvenation or terrace formation can be ascertained. Traverse or cross profiles of the valley of Iril river at different places (Fig. 5.18a, b, c, d, and e) present the nature of the valley from its source area to the lower valley section. Fig. 5.18a shows a gorge, where the depth nearly equals the width and is likely to develop a V - shaped valley, a characteristic feature of youthful stage having greater width measured at the top than the bottom (Fig. 5.18b).

Fig. 5.18c (near Saikul) and Fig. 5.18d (near Sagolmang) show a mature stage of the river where down cutting becomes almost imperceptibly slow and valley widening becomes dominant. This change in characteristic stream activity may be said to mark the transition from youth to maturity (Thornbury, 1985). Fig. 5.18e provides the character of a matured valley produced by the continued valley widening processes and forming wide open valley of old age which has a width considerably greater than that of the meander belt. Plate 5.1 shows the wide valley of Iril river at the confluence point with Imphal river at Lilong.

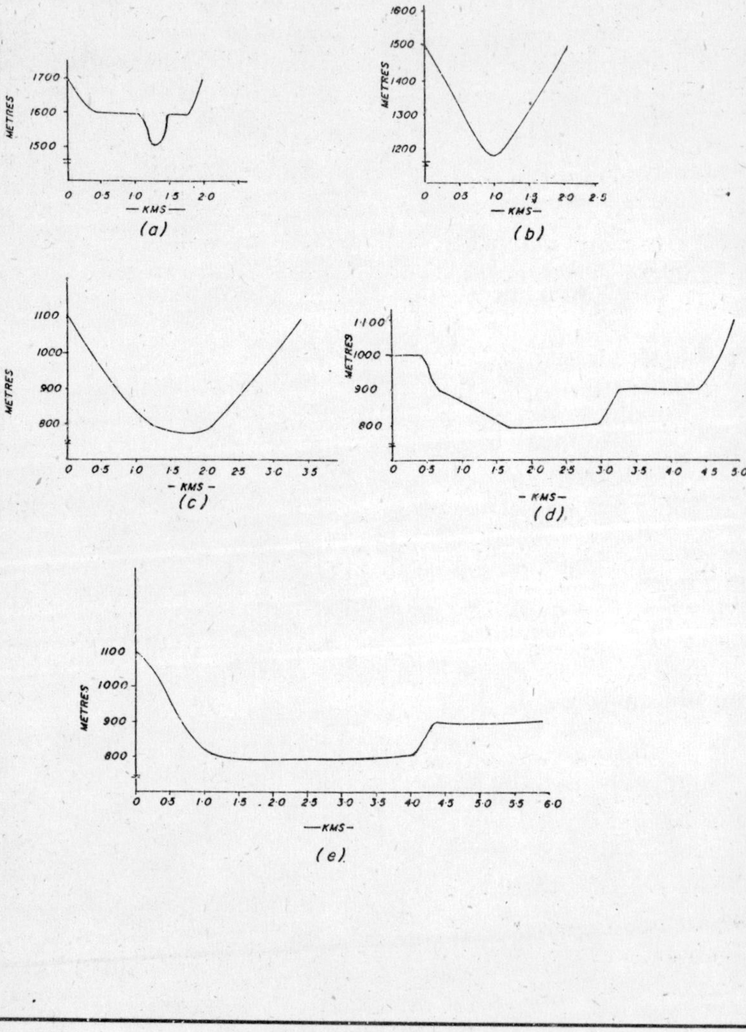

Fig. 5.18 a-e Traverse Valley Profile

5.16 RIVER TERRACES AND OTHER FEATURES OF STREAM DEPOSITION

The river terraces include the remnants of the former valley floor levels of the stream. Under the current cycle of fluvial erosion, the remnants of the valley floors stand a higher step in grades than the present river gradient. Plate 5.2 (Ichai Lok) shows the paired terraces used widely for cultivation purposes. Paired terraces again can be identified in plate 5.3 and one alluvial fan is seen in the right middle side of the plate. The village Ihang Karong is situated at the upper terrace region of the river which is also seen at the extreme left of the same photograph.

Plates 5.4 shows the channel bar near the confluence of Iril and Ithoi Lok. Alluvial fans are mostly developed in areas of sparse vegetation. In many places such alluvial fans are well developed in this basin. Meandering of the river course is another significant feature of the matured river. Plate 5.5 shows a big meandering course of the Iril river near Taretkhul.

5.17 CONCLUSION

From the analysis of slope, relief features, dissection index, restored or generalised contour, stream line surfaces, hypsometric curve, longitudinal profile, cross profile of the landscape, accordant of summit levels, traverse profile of the valley and other depositional features of the Iril river basin, it can be concluded that it is a matured basin which has experienced several cycles of erosion, transportation and deposition etc. Therefore, it may be regarded as a basin involved in multicyclic evolution of the landscape.

6

GEOMORPHIC REGIONS AND WATERSHED MANAGEMENT

An overall analysis of the foregoing chapters shows a distinct array of different geomorphic regions within the Iril river basin, which have their own characteristic morphology. Their morphological character, in turn, affects the overall nature of the landscape. On the basis of relative relief, intensity of relief, drainage density, drainage texture and dissection index analyses, Iril basin has been divided into five geomorphic units/landform regions. For these landform regions various models of landuse planning and watershed management have also been discussed in this section. The application part of the findings are supported by the landuse and maps showing potential area for land development and proposals for spatial planning.

6.1 LANDFORM REGIONS

Geomorphological map (Fig. 6.1) shows the overall geomorphological character of the Iril basin. Further, after correlating the results of various morphological studies, landform regions of the Iril basin have been outlined in Fig. 6.2, which reveals the different landform areas of the basin with reference to their morphological characters. The geomorphic map has been prepared on the basis of maps showing relative relief, intensity of relief, dissection index, drainage density, texture of dissection etc. All these maps have been compared with each other with outline of each categories of land surface portrayed according to their characteristic features. Table 6.1 shows the detailed characteristics of these geomorphic regions.

Table 6.1
Landform Region

Name of the units with morphometric characters	Total area	Area in per cent
A. High hill chains (Rh - Rhm, DIh, Df - Dm, Tf - Tm, Ih)	144.75	11.27
B. Dissected Hills		
(i) Highly dissected hills with scarp (Rh - Rhm, DIh - DIm, Df - Dm, Tf - Tm, Ih - Im).	338.7	26.36
(ii) Medium high dissected hills (Rm, DIm - DIl, Df - Dm, Ih - Im, Tf - Tm).	450.35	35.05
C. Lower valley region intermixed with residual hills (Rm - Rl, Im - Il, Dm, Dim - DIl, Im).	186.10	14.48
D. Lower plain region (Rl, DIl, Il, Dm - Dc, DIl).	165.10	12.84

Legends

1. Relative Relief (R) : Low (1), Below 1000 m ; moderate (m), 100 m - 300 m ; high (h), 300 m - 500 m and above.
2. Intensity of Relief (I) : Low (1), below 1000 m ; medium (m), 1000 m - 1600 m ; high (h), 1600 m - 2200 m and above.
3. Dissection Index (DI) : Low (1), 0.05 ; moderate (m) 0.05 - 0.20, high, (h) 0.20 - 0.40 and above.
4. Drainage Density (D) : Coarse (c), below 4 km/sq.km., medium (m), 4 - 12 km/sq.km.; fine (f), above 12 km/km2.
5. Drainage Texture (T) : Coarse (c), below 4/sq.km. ; medium (m), 8 - 12/sq.km. ; fine (f), above 12/sq.km..

Thus, the Iril basin comprises four broad geomorphic regions:

A. High hills

B. Dissected hills

C. Lower valley region intermixed with residual hills and

D. Lower plain region.

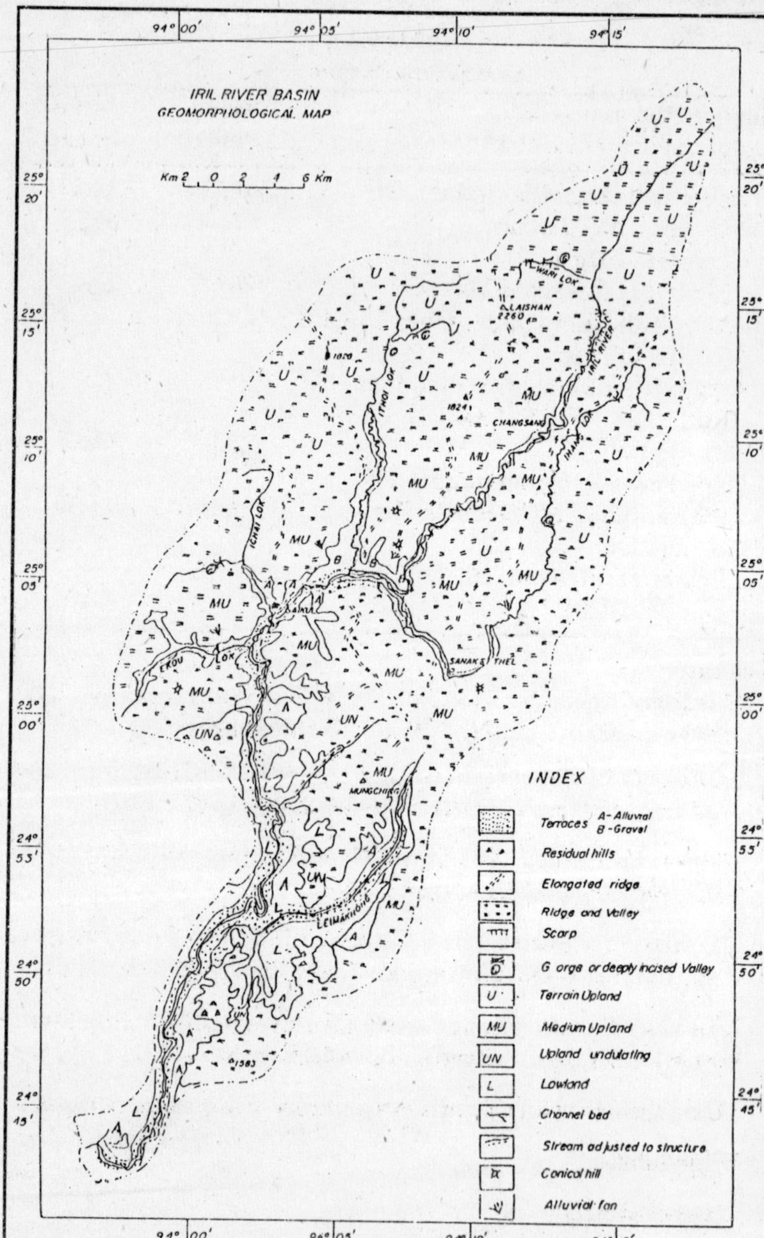
IRIL RIVER BASIN
GEOMORPHOLOGICAL MAP
Km 2 0 2 4 6 Km
94° 00'
94° 05'
94° 10'
94° 15'
25° 20'
25° 15'
25° 10'
25° 05'
25° 00'
24° 55'
24° 50'
24° 45'
LAISHAN 2260
WARI LOK
IRIL RIVER
ITHOI LOK
CHANGSANG
IHANG LOK
ICHAI LOK
SAIKUL
EKOU LOK
SANAKSI THEL
MUNGCHING
LEIMAKHONG
INDEX
Terraces A- Alluvial B -Gravel
Residual hills
Elongated ridge
Ridge and Valley
Scarp
G orge or deeply incised Valley
Terrain Upland
Medium Upland
Upland undulating
Lowland
Channel bed
Stream adjusted to structure
Conical hill
Alluvial fan

Fig. 6.1

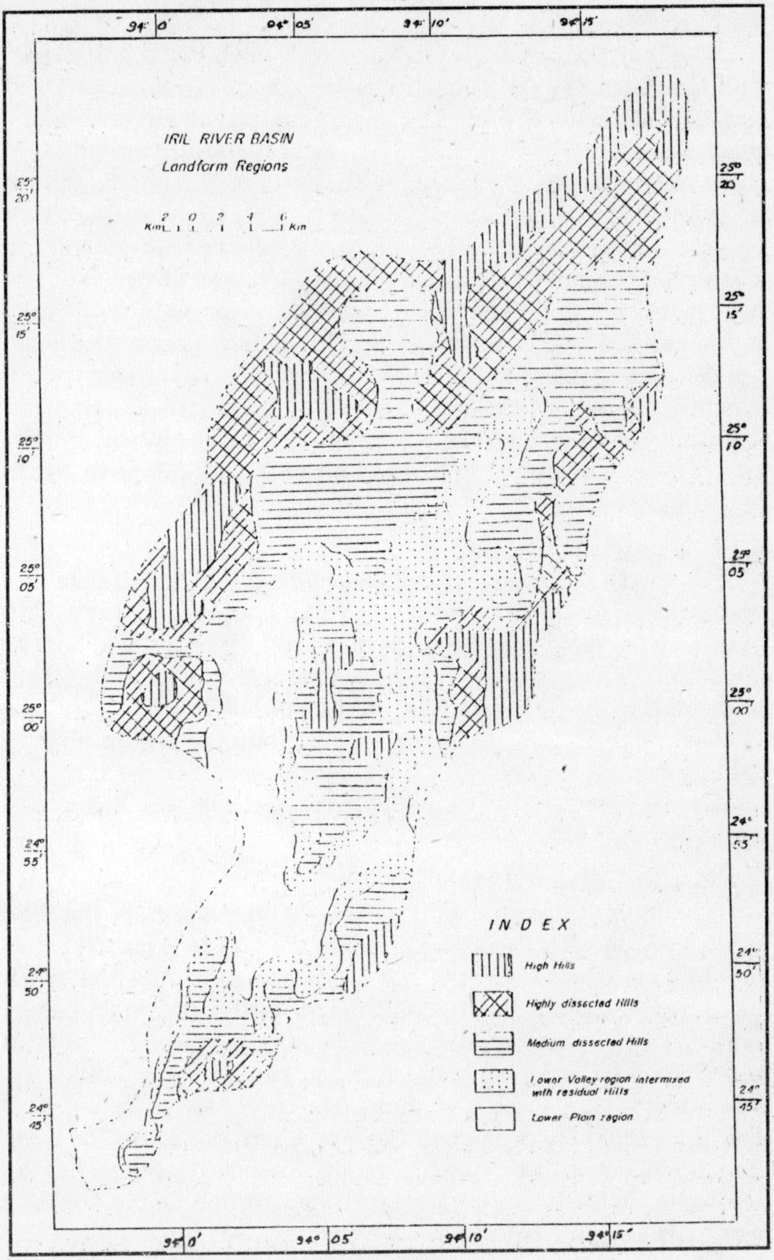

IRIL RIVER BASIN
Landform Regions
INDEX
High Hills
Highly dissected Hills
Medium dissected Hills
Lower Valley region intermixed with residual Hills
Lower Plain region

Fig. 6.2

6.1.1 HIGH HILLS

The high hills occupy 11.27 per cent of the total basin area. This landform region is more extensive in the northern part and scattered along the border as patches of high relief. Higher relative relief, fine to medium texture of dissection and high to medium dissection index are associated with this type of landform. From the point of agriculture and settlement, this region is not important because agriculture is possible only on the hill slopes where shifting cultivation *(jhuming)* is normally practised. *Jhum* cultivation causes heavy destruction to the forest cover degrading the forest area into scrub land. This region forms the highest part of the basin area including the highest peak at 2473 m (source area of the Iril river). These hill slopes are dissected by numerous streams and tributaries of the Iril river. Intensity of relief in the high hills varies from 2200 m to 1600 m and relative relief from 300 m to 500 m and above.

6.1.2 DISSECTED HILLS

The dissected hills cover an area of 61.41 per cent of the basin representing the biggest surface coverage of the landscape. This region is characterised by high hills well dissected by various streamlets. Relative relief in this part varies from 100 m to 300 m with high to medium intensity of slope. Drainage texture and drainage density are quite high. This landform region may be grouped into two divisions on the basis of its dissection and landform characteristics : a. highly dissected hills, and b. medium high dissected hills.

a. Highly Dissected Hills

The highly dissected hills occupy 26.36 per cent of the total basin area. Next to the high hill belt in the northern part, it forms an elongated belt from the north eastern side upto the south western side dotted by high hills in the north western and some portion of the south western section. In this region, intensity of relief varies from 500 m to 200 m per sq km. Dissection index too varies from 0.21 to over 0.41 showing highly dissected relief. This landform region occupies the upper northern portion of the basin and tapers towards the south western and south eastern parts of the basin. It further occupies an elongated belt in the south to south eastern parts of the basin. Drainage density greatly fluctuates between 2km /sq km and 6km /sq km and texture varies from fine to medium (4 to 12, sq km) in this section, thus showing an area of high dissection index.

b. Medium High Dissected Hills

This landform region covers over one - third, more precisely 35.05 per cent of the total basin area. The medium high dissected hills form the middle upper portion of the basin and comprise a belt running from south to south eastern part of the area. Dissection index in the region varies from 0.06 to 0.40 with dominance of the dissection index 0.11 to 0.20. The drainage texture also varies from high to medium sometimes unevenly located coarse texture. Though relative relief varies from 100m to 300m it is dominated by 100m - 200m region. This landform region, lying in the eastern part of the basin, is small and less extensive.

6.1.3 LOWER VALLEY REGION INTERMIXED WITH RESIDUAL HILLS

This landform region is found in the central part of the basin enclosed by the medium high dissected hills in the northern and eastern parts. It occupies only 14.48 per cent of the total basin area. The relative relief varies from medium to low, with dominance of medium relative relief covered by 100m to 300m level. Intensity of relief is almost low but dotted by medium intensity slope areas. Drainage texture is almost medium (8 - 12 per sq km). However in some areas fine and coarse drainage texture are also found. Dissection index shows low to medium, with dominance of low dissection index.

6.1.4 LOWER PLAIN REGION

Lower plain region, forming the low relief tract (100m and below), covers 12.84 per cent of the total basin area. Intensity of slope is the least in the region. Although in some parts medium drainage density is observed, as a whole, this section is characterized with low drainage density. Dissection index varies from 0.30 to 0, thus making it a low dissection index region. However, in the upper or the north western part of the basin, the dissection index varies from 0.30 to 0.20. Texture of dissection is also coarse showing 0 - 4 per sq km. This has resulted in the lowest dissection index (below 0.05 per sq km) for the region. Although this landform region is small, it has a significant role in agriculture as well as settlement. Most of this region is thickly populated and provides a sound base for agricultural activities.

The study of the landform regions of the Iril river basin defines the morphological personality of each of them. Therefore, analysis of these landform regions can further lead to a proper utilisation of its land resources.

Table 6.2
Land use pattern

Category	Area in sq km	% of the total basin
1. FOREST		
a. Degraded forest / scrub land	662.25	51.53
b. Forest blank due to heavy deforestation and grazing	21.42	1.67
2. CULTIVABLE LAND		
a. Shifting cultivation	413.86	32.21
b. Crop land in valley region	186.69	14.53
3. BUILT UP LAND	0.78	0.06

6.2 LAND USE PATTERN

For a developing economy, proper and optimum utilisation of the natural resources, particularly its land and water resources, is very essential. Proper utilisation of the land and water resources strictly comes under the utilisation of watershed, which strictly refers to the divide separating one drainage basin from another.

While studying the land resources of the Iril basin the study of its land, soil characteristics such as physiography, slope, nature of the soil, present land use, geology, climate, socio - economy, legal aspects, hydrological behaviours etc. are imperative for evaluation of the most productive interaction between land and water resources. As the land use is essentially under the human control, it can be effectively planned as part of the drainage basin management. Table 6.2 shows the land utilisation pattern in the Iril basin.

6.2.1 FOREST

Degraded scrub forests extensively cover the watershed of

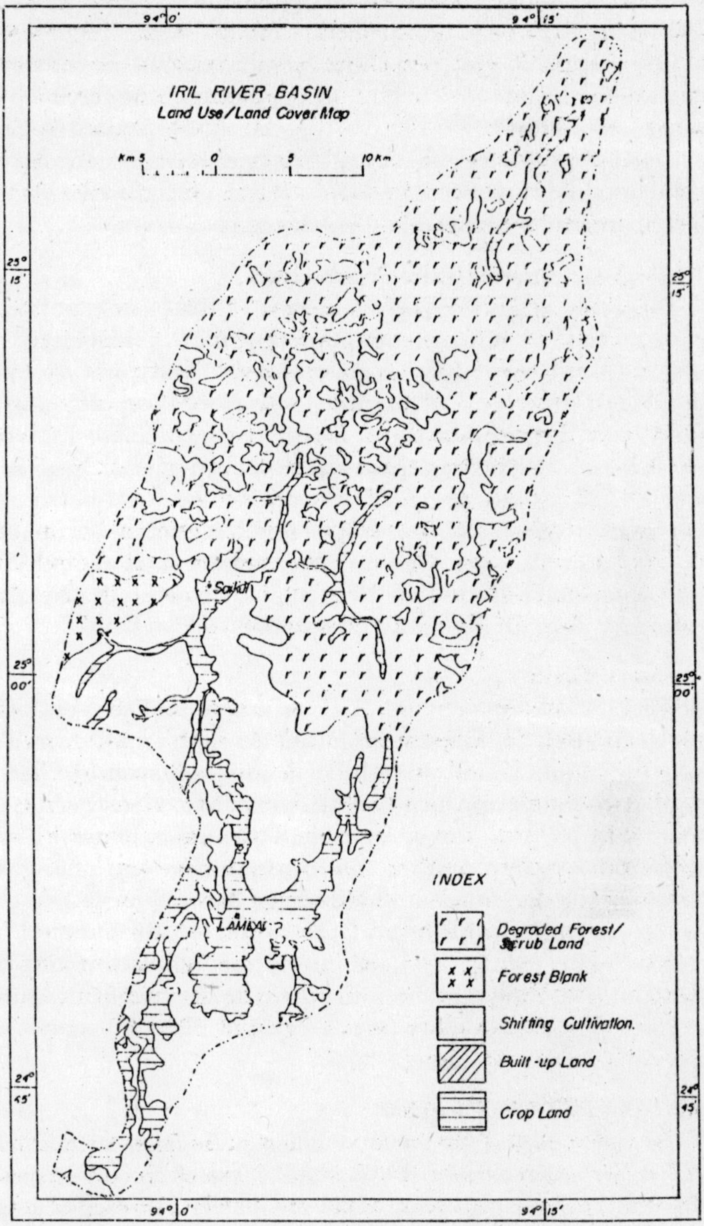

IRIL RIVER BASIN
Land Use/Land Cover Map
Km 5 0 5 10 Km
94° 0'
94° 15'
25° 15'
25° 00'
24° 45'
Saikul
LAMLAI
INDEX
Degraded Forest/ Scrub Land
Forest Blank
Shifting Cultivation.
Built-up Land
Crop Land

Fig. 6.3

the Iril river. Natural forests, covering the northern upper reaches of the basin, have been diminished to much extent. Uncontrolled and unscientific logging and shifting cultivation in the catchment area are responsible for such rapid depletion of the forest cover. The protective and productive forests of the basin are being destroyed in preference to a very brief period of immediate crop production under *jhum* cultivation system, which further leads to a wide spread denudation and enhanced soil losses.

a. Degraded Forest or the Scrub Land

This type of land covers most part of the basin, spotted by areas affected by shifting cultivation (Fig. 6.3). Scrubs have been resulted from clear felling on steep slopes. Faulty and excessive logging and firewood cutting have adversely been affecting the forest floors during extraction. This can be seen in Plate 6.1, where the whole hillock has been changed into a barren land. Scrub lands cover 662.25 sq km or 51.53 per cent of the total basin area. Overgrazing is basically responsible for changing the scrub forest into barren landscapes. Geomorphic regions such as high hills, highly dissected hills and medium high dissected hills are almost covered by degraded forest with patches of *jhum* lands.

b. Forest Bank

Forest blank area covers 21.42 sq km or 1.67 per cent of the total basin area. Such areas are found along the road from Ekou Bazar to Saikul. Totally bare hillocks are well seen in Plate 6.2 showing the landscape near Saikul. Forest fires affect the hills and make them barren. Forest fire results in the extensive loss of vegetation, organic matter, micro organisms and changes in hydrological behaviours of the drainage basin. Forests are set on fire for *jhuming*, hunting and for collection of charcoal and firewood. The young trees are gutted for defoliation and later felled. Since nothing remains other than the short charred stumps, the forest becomes a blank area overnight. Plate 6.3 shows such an area near Natum Ching.

6.2.2 CULTIVABLE LAND

As major part of the basin area lies in the hills, the cultivable land in the valley section is too small. Most of the hill slopes are subjected to dry cultivation while the levelled stretch of land in the valley is used for wet cultivation. Thus, cultivable lands are of two types : a. cultivable land in the hilly region, and b. cultivable land in the valley region.

a. Cultivable Land in the Hilly Region

The cultivable land in the hilly region covers 413.86 sq km or 32.21 per cent of the entire drainage basin. However, it has been fluctuating from time to time with shifting of the fields under *jhum* cultivation. This type of cultivation is associated with forest fire also. The forests are burnt and cleared before manually digging them for growing paddy, maize, pumpkin and other vegetables. The plots used for *jhum* cultivation for two or three consecutive years are abandoned due to exhaustion of nutrients. The abandoned plots are allowed to regenerate for a few years and put to reuse. This type of cultivation is known as "shifting cultivation" or *jhuming*. The people will shift their *jhum* fields from one slope to another and continue the same process, thus causing soil erosion and siltation in the drainage basin.

The other type of cultivation practised in the region is tarrace cultivation, which is not so common in the high hill areas. Most of the hill people resort to shifting cultivation for their food requirements and livelihood.

b. Cultivable Land in the Valley Region

Cultivable land in the valley portion of the basin is very little. Only 186.69 sq km or 14.53 per cent of the total basin area is under this type of land use. Broad river beds, river terraces and valley lands are used for sedentary cultivation. For the absence of irrigation facilities, monsoon rains provide the only sources of water to the cultivated land, which remains fallow and least used during dry seasons. Both *jhum* cultivation and wet valley cultivation areas are clearly seen in Plate 6.4 depicting parts of the area near Saikul.

Other types of land utilisation are also found along the Iril river. The settlements such as Sagolmang, Tiger Camp, Kangla Siphai, Kangla Sangomsang, Taretkhul, Moirang Kampu, Phaknung, Naharup, Khetrigao, Irilbung, Urup, Arapti and Lilong etc. are situated on both the banks of the Iril river. River terraces near these villages are utilised for plantation of crops such as pea, cowpea, mustard, cabbage etc. during winter season (Plate 6.5). But near the villages Naharup, Urup, Arapti and Lilong (confluence point with Imphal river) the river banks are planted with bamboo groves to meet the demand of rural house - building material and to check soil erosion and destruction of the river banks during floods in rainy season (Plate 6.6).

6.2.3 BUILT UP LAND

Built up land covers only 0.78 sq km or 0.06 per cent of the total basin area. This shows that this drainage basin is least developed with very little urbanisation and poor economic growth.

6.3 LANDUSE PLANNING AND WATERSHED MANAGEMENT

In the foregoing paragraphs an attempt has been made to analyse the existing landform characteristics and present land use pattern. The various limitations and impediments in the utilisation of land resources are all related to physiography, rainfall characteristics and cultural factors. The watershed approach can be integrated with programmes like forestry, soil conservation, rural and community development, and farming system (NIRD, 1989). It is a well known fact that land and water interact to produce food, fodder, firewood, timber and raw materials for the people, their domestic animals and agrobased industries. Therefore, a well planned landuse strategy is essential for obtaining the most productive interaction between these resources.

The following suggestions have been made for proper landuse management in the Iril river basin.

6.3.1 PLANNING STRATEGY

It is necessary to solve the problems that arise while planning the landuse in the Iril watershed and suggest plans to minimize the misuse of land resources. Singh (1970) proposed that "the landuse should be re-oriented with a flexible and dynamic plan aimed at providing maximum utility for every acre of land". However, in this context land fragmentation is one of the major problems in this region. Land use on private land cannot be influenced by mere legislation. There are other drawbacks in planning, *viz.* lack of communication facility, financial support, proper understanding etc. For example, in the hilly region if the land is recommended for fruit production in lieu of food production, it cannot materialise due to lack of financial support to cater consumption expenses during intervening period etc. Therefore, the problem of landuse changes should be tackled through price policies, credit / subsidies, incentives and other possible helps. Therefore, farmers are to be fully involved, encouraged and assisted on a massive scale.

Since watershed management demands full understanding of the ecological endowments and physical as well as anthropogenic limitations for a comprehensive treatment of the target area, it is difficult to suggest measures at this stage particularly in absence of a systematic data of soil survey and land capability analysis. The existing settlement areas and communication facilities in the Iril basin are shown in Fig. 6.4. Based on the information derived from the present study of landforms, an integrated approach comprising the following elements can be suggested for treatment of the different watersheds in the Iril river basin, *viz.*, a. Treatment of forest lands, b. Treatment of agricultural lands, c. Treatment of wastelands, d. Soil and water conservation, e. Development of subsidiary occupation, and f. Community development programmes

a. Treatment of Forest Lands

The sub - tropical wet hill forests used to occupy the height range of 900 - 1800m in earlier times. In the higher reaches the wet temperate forests were found but in a very limited area. However, various man induced stresses like *jhuming*, wildfire and indescriminate felling of trees for charcoal, firewood and timber have converted these forests into scrub forests which occupy 51.53 per cent of the total basin area at present.

The "Village Fire Reserves" are the luxuriant tree forests maintained by tribal people in the immediate surroundings of their villages atop hills as protection belts. This can be converted into a large scale conservation practice with the application of scientific selective felling of trees. Thus, in the higher reaches (above 1800m) commercial timber species such as Michelia, Phoebe, Lagerstroemia, and Gmelina can be grown. In the lower regions (900 - 1800m) both timber species like Pine can be mixed with firewood species like Quercus sp., and Bauhinia sp. and Lista polianata etc.. No reliable means of reducing the firewood and timber demand has yet been introduced in the state. Therefore, rehabilitation of forests in the region can be achieved through, a. a large scale afforestation programme, and b. a ban on indiscriminate logging by contractors and burning the forest for

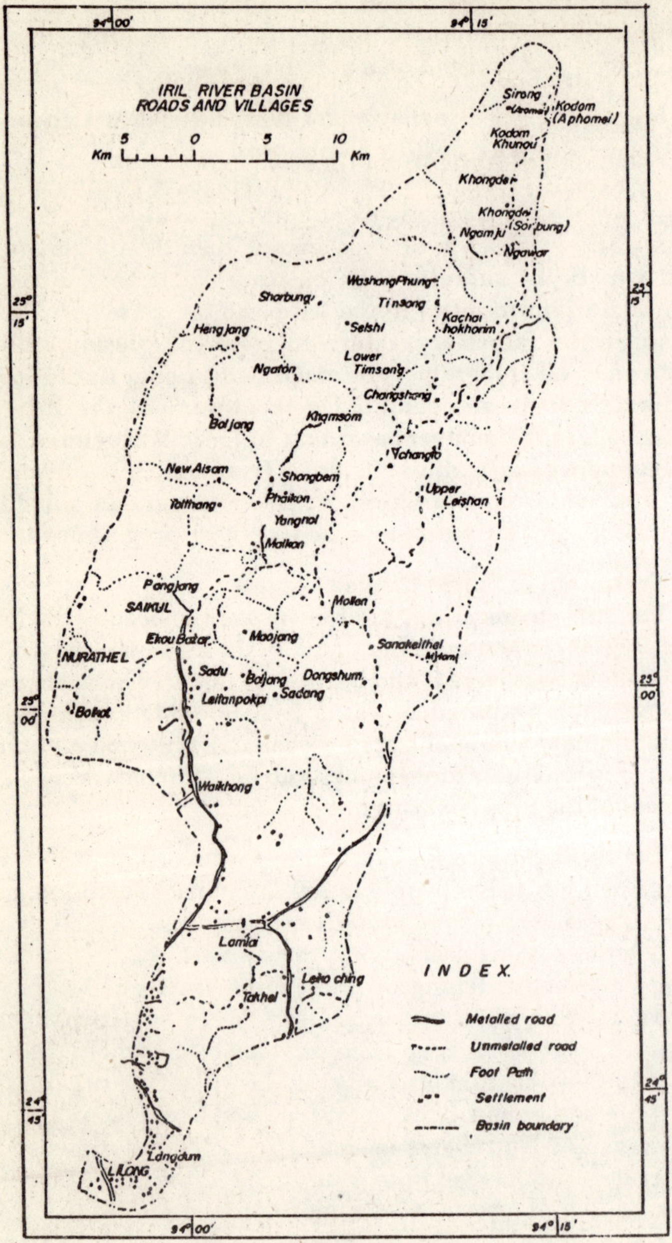
IRIL RIVER BASIN
ROADS AND VILLAGES
5 0 5 10
Km Km
94° 00'
94° 15'
25° 15'
25° 00'
24° 45'
Sirong
Kodom (Aphomei)
Kodom Khunou
Khongdei
Khongdei (Sorbung)
Ngamju
Ngawar
WashongPhung
Sharbung
Tinsong
Selshi
Kachai hokhorim
Hengjang
Ngaton
Lower Timsong
Changshang
Baljang
Khamsom
Ychangio
New Aisam
Shangbem
Upper Leishan
Totthang
Phaikon
Yangnoi
Maikon
Pangjang
SAIKUL
Mollen
Ekou Bazar
Maojang
Sanakeithel
NURATHEL
Sadu
Baljang
Dongshum
Leitanpokpi
Sadang
Bolkot
Waikhong
Lamlai
Leiko ching
Takhel
Langdum
LILONG
INDEX
Metalled road
Unmetalled road
Foot Path
Settlement
Basin boundary

Fig. 6.4

shifting cultivation.

Altogether 388.83 sq km of higher lands with slopes above 17° are available in the Iril catchment for rehabilitation of forests.

b. Treatment of Agricultural Lands

Agricultural lands in the present context can be grouped into two for the purpose of proper treatment, *viz.*, i. Upland cultivated lands, and ii. lowland cultivated lands

i. Upland Cultivated Lands

The unremunerative and harmful practice of *jhum* cultivation (locally called *Pamlou*) is found prevalent over the hill slopes with gradient of 7° to 17°, shown in Fig. 6.5 as areas of second and third preference for land development. Altogether an area of 496.25 sq km having been seriously affected by deforestation is available for treatment through changes in agricultural practices. This zone of upland cultivation can be put to two kinds of treatment *viz.*

(a) Silvi - pastoral and horticultural zone occupying the upper regions between 11° and 17° slope. This region covering an area of 216.48 sq km is just below the tree forest zone. In this region, fuel and fodder species of trees can be planted along with horticultural crops, such as,

Fruits : orange, lemon, cashewnut, pineapple, plum and banana.

Cash crops : rubber, coffee, cardamom and citronella.

Vegetables and spices : potato, turmeric and ginger.

Halfmoon terraces can be constructed whenever and wherever necessary in this horticultural zone.

(b) Terraced agricultural zone occupies the lower hill slopes between the 7° and 10° gradient. An area of 279.77 sq km is available to this kind of land treatment. Various structures like contour bunds, bench terraces, ridge terraces and grassed waterways can be constructed to conserve water and soil required for the upland wet cultivation in this part of the basin. A number of crops can be grown sucessfully in this zone, *viz.*, maize, ragi,

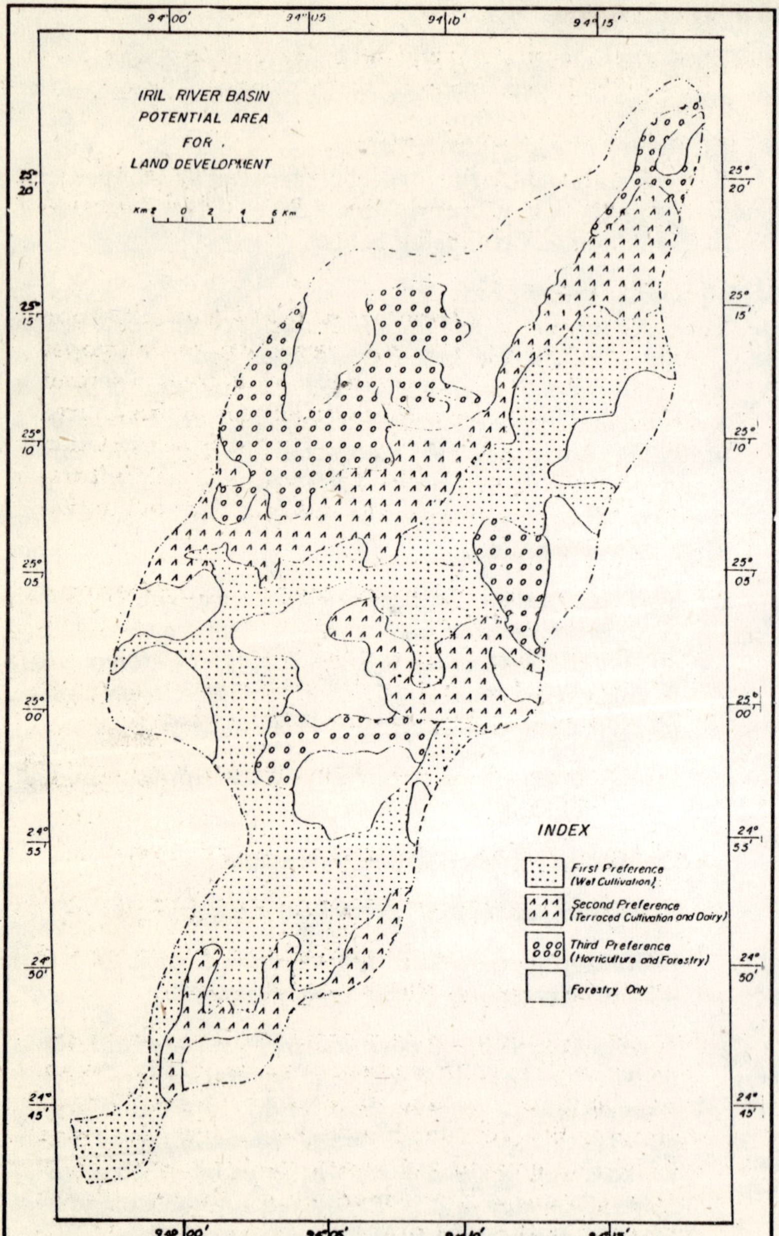
IRIL RIVER BASIN
POTENTIAL AREA
FOR
LAND DEVELOPMENT
Km 2 0 2 4 6 Km
INDEX
First Preference
(Wet Cultivation)
Second Preference
(Terraced Cultivation and Dairy)
Third Preference
(Horticulture and Forestry)
Forestry Only
94° 00'
94° 05'
94° 10'
94° 15'
25° 20'
25° 15'
25° 10'
25° 05'
25° 00'
24° 55'
24° 50'
24° 45'

Fig. 6.5

tapioca, groundnut, arhar, rice - bean, soyabean and potato in the upper terraces, rice in the lower terraces and, hybrid napier grasses on the risers for fodder and reeds etc. along the waterways for thatching.

Wherever the terracing is not possible, improved *jhuming* with pedigree seeds, improvised tools and with application of manures and fertilizers can be practised.

ii. *Lowland Cultivated Lands*
The lowland cultivated lands are found on both sides of the Iril river in its meandering course in the valley where the average slope ranges betwen 0^0 and 6^0. The total area available in this zone is 399.02 sq km. This is the zone of first preference and is much more productive than the upland *jhum* cultivated tracts. Major clusters of settlements are also found in this zone. However, these wet cultivated lands are rainfed and remain fallow during winters as there is hardly any irrigation facility. The discharges of major tributaries in the plain areas like Ekou, Khamen Lok and Leimakhong may be examined and pick up weirs constructed on them at the foothills for development of minor irrigation. Such an arrangement can convert a sizable portion of the single cropped rice fields into multi - cropped productive fields. The modern scientific agricultural inputs like HYV seeds, fertilisers and pesticides, machanisation to a limited extent and crop rotation are suggested for this type of land.

c. Treatment of Wastelands
The watersheds occuring on the Iril river basin belong to three categories, *viz.*, i. The denuded rilled and gullied unproductive lands on degraded hill slopes with shallow soil, ii. The undeveloped lands at the foothills (piedmont zone), and iii. The waterlogged areas found in the sediment filled wetlands like the Phaknung pat.

Treatment of these wastelands can be brought about in different ways.

i. *The Degraded Hill Slopes*
These are mostly the abandoned *jhum*lands which in earlier times got rejuvenated during 20 - 30 years of *jhum* cycle. Since, the *jhum* cycle has been reduced to 3 - 5 years, their natural

course of top soil replenishment has to be assisted by various upland cultivation practices as described ealier.

ii. *The Foothills*

Foothills are found often left uncultivated. Foothills are the bridge between the hill cultivation and the valley cultivation. These can profitably be reclaimed for cultivation of crops like sugarcane, cotton, vegetables and pulses. Foothills are the best places for development of pastures and dairy farming.

iii. *The Waterlogged Areas*

The Phaknung pat is a derelict wetland. It is no more in the shape of a lake. It cannot be used for cultivation of rice either. Therefore, it is suggested that the wetland may be put to multipurpose cooperative farming including : pisciculture, cultivation of high water tolerant varieties of rice, cultivation of aquatic food plants and reeds like, Euryale forex (*Thangjing*), Nelumbo spp. (*Thambal*), Neptunia prostata (*Ishing Ikaithabi*), Zizania caduciflora (*Ishing Kambong*), Scirpus spp. (*Kouna*), and Sagittaria spp. (*Koukha*)

d. Soil and Water Conservation

The toposequence of different landuse patterns *i.e.* tree forests - silvi - pastoral - horticultural - terraced cultivation from top to the bottom along the hill slopes as suggested will help in the conservation of soil and water resources as this will check the sheet erosion of the hill slopes and permit rain water to percolate into the soil. As additional measures, green manures can be put while a simultaneous cultivation of nitrogen fixing legumious plants can enrich the soil quality. Clove is a well proven plant for this purpose. Besides, various mechanical structures like silt retention dams, *e.g.* vegetative check dams, logwood check dams, stone rubble check dams and masonary check dams can be constructed wherever there is intensive erosion.

Storage of runoff water in larger cisterns downhills is also helpful in conservation of water.

Since a drastic control on cattle grazing effect on hill slopes is needed to protect the soil cover, an efficient range management system has to be evolved to develop green pastures in the piedmont zone.

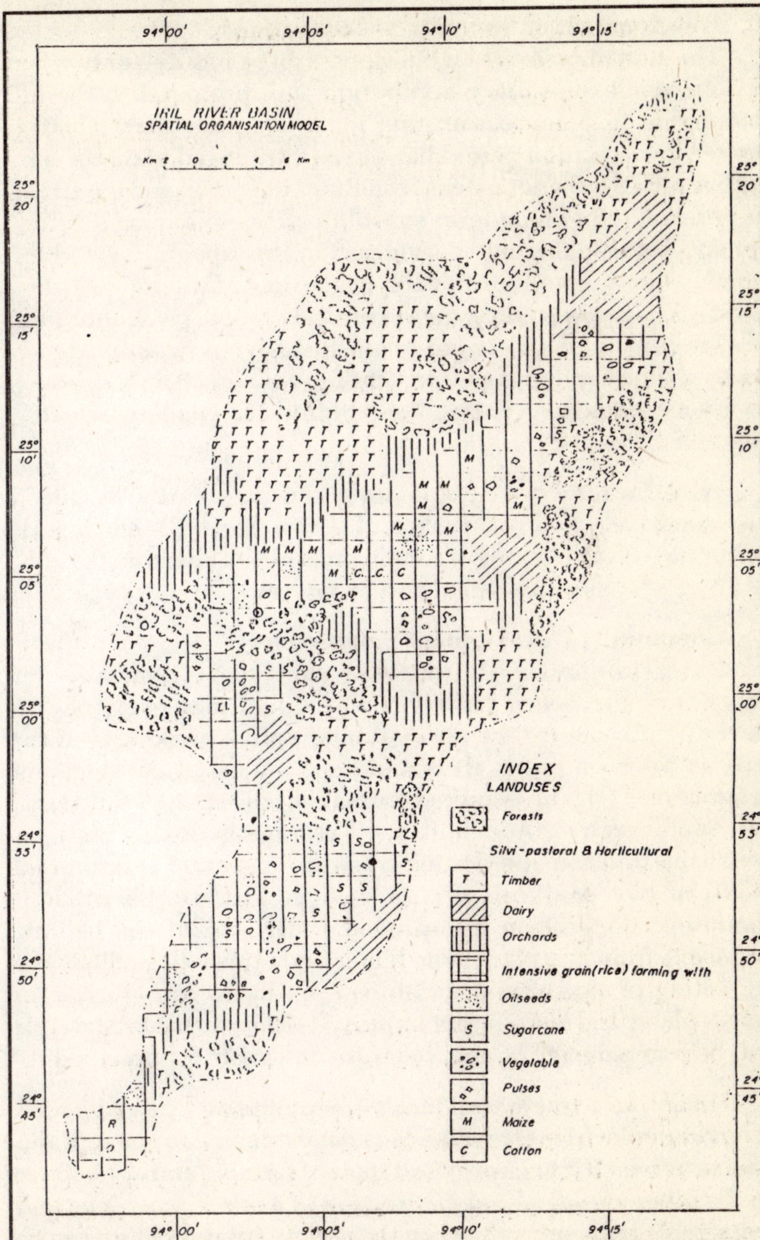
IRIL RIVER BASIN
SPATIAL ORGANISATION MODEL
Km 2 0 2 4 6 Km
94°00'
94°05'
94°10'
94°15'
25° 20'
25° 15'
25° 10'
25° 05'
25° 00'
24° 55'
24° 50'
24° 45'
INDEX
LANDUSES
Forests
Silvi-pastoral & Horticultural
Timber
Dairy
Orchards
Intensive grain(rice) farming with
Oilseeds
Sugarcane
Vegetable
Pulses
Maize
Cotton

Fig. 6.6

e. Development of Subsidiary Occupations

The human pressure on hill slopes can be reduced only when certain viable subsidiary occupations are provided to the hill people for sustenance. Farming methods other than *jhuming*, having low gestation period have to be introduced to fill the time lag between the abandonment of shifting cultivation and the arrival or return of other long term substitutes. The following schemes are suggested taking into consideration the topography and the agro - climatic conditions of the Iril basin (i) farming including livestock, piggery, poultry, mushroom cultivation, pisciculture and beekeeping, and (ii) small scale industries such as dairy, particularly cheese and *ghee* processing, cane and bamboo works, fruit juice extraction centres, dehydration and canning industry, rice mills.(Fig. 6.6).

While the farming activities can be carried out in the reclaimed piedmont areas and other areas of gentle slope, the small scale industries can be planned at certain settlement clusters in the zone of first preference like Saikul and Sagolmang.

f. Community Development Programmes

Spatial integration of all activities suggested is imperative for an overall success of Iril river watershed management. An elaborate scheme for the promotion of agro - processing in the area as has been put in the small scale industrial development programmes is highly required for making all activities functional and remunerative. Adequate food storage godowns both for perishable and non - perishable crops are suggested at important clusters like Malkon village and Saikul. If the present communication system is improved a little, Saikul will become accessible from any place in its hinterland. This will facilitate the marketing of agricultural and forest produces. Introduction of biogas plants and fuel efficient improved *chulhas* to the local people will help in reducing human pressure on forests for firewood.

Finally, an active mass education programme for assimilation of environment friendly methods of development on one side and a hectic research into improvised tools of farming and agriculture on the other are suggested for creation of a self regenerating and sustainable economic growth in the region. Iril river basin can be a model watershed for adoption of scientific management for environment sound economic development.

7

SUMMARY AND CONCLUSION

The Iril river basin lies in the north eastern part of the Manipur state. It extents between latitude 24°40' N - 25°25'N and longitudes 93°55'E - 94°20'E covering an area of 1,285 sq km. The basin lies within the administrative districts of Ukhrul, Senapati, Imphal East and Thoubal. Iril river originates from the south eastern slope of Hougdu Ching range in Ukhrul district at a height of 2473m above mean sea level and flows through the hill tracts of Ukhrul and Senapati districts and descends to the almost levelled tracts of Imphal East and Thoubal districts in the Manipur valley. It debouches its water into Manipur river (known as Imphal river in its upper part before its confluence with Khuga river), a tributary of the Chindwin - Irrawaddy drainage. Although Iril river basin is a sub - basin of the Chindwin - Irrawaddy system, in the present context it is termed as Iril basin accepting the drainage area as a distinct geomorphic unit. Iril basin is marked by different features of erosional and depositional origin, fluvial landforms and many other interesting geomorphological features relating to the evolution of a variegated polycyclic landscape.

Physiographically the basin is characterised by alternate hills and valleys having sharp crested ridges running NNE - SSW to NE - SW direction. The river is 144.5 km long and flows almost in the middle of the basin. It receives a number of tributaries, some of them being considerably long *viz.* Ihang Lok, Ithoi Lok, Ichai Lok, Ekou Lok, Khamen Lok, Leimakhong and Wari Lok etc.

The Iril basin can be divided into two distinct parts - the hill section and the valley portion. The greater part of the drainage basin lies in the hill section. Geologically the rocks of the basin area belong to formations of different geologic ages *viz.* Recent, Barail (upper Eocene to Oligocene) and Disang (upper Cretaceous to Eocene) The rock types associated with Recent group are alluvium, pebbles and fine soils covering the valley portions. The

Barail group of rocks are predominantly arenaceous in character with thick beds of sandstone, shale and siltstone mostly occurring at the core of the major synclines with gradational contact and marks of coarse gritty sandstone. The rock types, found in Disang group, are somewhat argillaceous in character with predominance of shale and siltstone. Siltstone beds are relatively not common. The siltstone found in the basin is generally soft and friable in nature. Major part of the basin is covered by the rocks of Disang formation.

This area has a complex history of geological evolution which is indicated by multiple folding, faulting and shearing. Three different phases of folding are observed in the Iril basin area. The first generation of folding was in the form of localised tight isoclinal to open type of folds with NE-SW trending sub-vertical areal planes and low to moderate plunges. The second generation of folds were in the form of regional anticlines and synclines affecting the formations and showing various types of folding from open asymetrical to tight symmetrical folds. The third generation of folds came in the form of broad warping, trending almost E-W, characterised by prominent and persistent upstanding feature of sandstone and siltstone in Mungching and the areas north, NE, NW and east of it. Many of the major strike faults have restricted the domain of different phases of folding and have also controlled the topography and drainage of the basin area. The middle part of the Iril river itself is controlled by major fault *i.e.* the Iril fault. Some of its tributaries too flow along the fault lines.

The type of index microfossils found in the basin also support the presence of Barail and Disang group of rocks. Landforms such as cliffs, alluvial fans, terraces and erosional surfaces are the characteristic features of the basin which are encountered in course of its downward movement. The valley portion is mostly covered by fertile alluvium, which contains deep silts deposited by the Iril and its tributaries.

Climatically this basin area enjoys a humid to subtropical monsoon climate. However, this climate largely prevails in the relief region upto 1800m, and the higher regions enjoy more or less a temperate climate. Annual rainfall is sometime as low as 837 mm and sometime as high as 2047 mm and temperature ranges between 1.5^0 and 37^0C.

The main soil types found in the basin area are the ferrugineous red gravelly soil which belongs to the non - laterized red soil (*Ultisol*) group containing a good amount of nitrogen phosphate and potash, and the *Entisol*, formed of the materials deposited by river Iril and her tributaries, having grey to pale brown colour and containing a good proportion of potash and phosphate and moderate quantities of nitrogen and organic matter.

The Iril river basin is manifested with drainage patterns like dendritic to sub - dendritic, rectangular or angulate and radial. The river shows three different channel behaviours that is, the upper portion where its course is straight, the middle portion which is angulate in nature, and the lower plain area showing meandering courses. Although dendritic pattern generally associates with homogeneous rock type with little adjustment to underlying structure, in this basin some of the sub - tributaries show more or less straight courses which can be explained as the effect of structural control. Further, as the basin area is typified by folded structures, where evidences of superimposition are found, presence of dendritic pattern is natural. Rectangular or angulate patterns are generally associated with underlying structures, specially of the attributes of joints or faults. Therefore, it may be stated that the course of Iril river is partly controlled by the underlying geologic structure. Radial type of drainage pattern expresses erosional slope of the region. This type of drainage pattern is found at Nongmaiching, south western part of Sanakeithel and western and south eastern part of Saikul area, thus showing the presence of many erosional hill slopes/surfaces within the basin.

As the study of drainage basin reflects the geomorphic features and evolution of landforms, analyses of parameters such as number of streams, number of confluences, length of stream, bifurcation ratio, drainage density, stream frequency and basin area have been made in the present study.

The maps showing number of confluence per sq.km. by using the choropleth and isopleth methods depict the density of confluences expressing three density categories. The highest number is over ten whereas the minimum is less than two. The highest confluence density is found along the south eastern and north western portions of the basin area. The minimum number of confluence covers 65.40 per cent, moderate 30.09 per cent and high 4.51 per cent of the total basin area. The hydrographic maps, *i.e.*, maps showing number of streams per sq. km. by choropleth and isopleth techniques, divide the basin area into three categories of the drainage texture - fine, medium and coarse covering 1.11 per cent, 66.91 per cent and 31.98 per cent of the basin area respectively. The coarse and medium texture areas are mostly confined to the southern and central portions of the basin. Fine draiange texture corresponds to the area of high relief region. As such, coarse drainage texture reveals the area of low relief. In highly dissected areas the number of streams is higher than that of the confluence points.

The studies about number of confluence and number of streams also help in analysing the texture of dissection, expressed as Thalweg length divided by area, mainly based on the length of streams. Dissection may differ in two areas having the same drainage density for the greater number of streams in one part or the other. Three categories such as fine, medium and coarse index of texture of dissection are defined from the length of stream with reference to its intervening spaces which vary in degrees. The fine or highly dissected areas are almost scattered but confine to some portions of the middle and the south western part of the basin. They cover 12.38 per cent of the total basin area showing texture of dissection index 4 to 6 km thalweg length per kilometre. The medium texture covers 59.14 per cent of the total area, which represent the valley section of the river and the lower relief tracts of the basin.

After comparing the drainage density map and the map for number of streams we can conclude that, (i) the highly dissected area has 12-16 stream per sq km or 4.6 Thalweg length per km, indicating high drainage density; (ii) in dissected areas both minor and major streams are very distinct and the total length of minor streams exceeds the total length of major streams per unit area, thus indicating highly dissected area or the area having fine drainage texture. Drainage texture is also regulated by the rate of rainfall and permeability of the mantle rock. This basin receives enough annual rainfall to run off quickly, thus creating a lot of drainage lines.

By analysing the different properties of the drainage basin we visualize a clear picture of the basin in connection with the degree of the adjustment among the left and right bank tributaries. Some of the left and right bank tributaries, *viz*. Ithoi, Ihang, Ichai, Wari, Leimakhong and Ekou have been taken into account to analyse their morphometric characters for the understanding of the present and past forms of the landscape. In these basins Ihang (196.25 sq km), Ithoi (237.50 sq km) and Leimakhong (112.03 sq km) are the 6th order streams and Ekou (56.14 sq km), Ichai (69.02 sq km) and Wari (34.97 sq km) are the 5th order streams. By analysing the bifurcation ratios from the above drainage basins, it is observed that the average bifurcation value of this basin conforms to the values calculated by Horton and Strahler. Bifurcation ratios characteristically range between 3.0 to 5.0 for watersheds in which the geologic structures do not distort the drainage pattern. However, in this region, although the average value of each of the stream does not exceed the value calculated by Horton,. some of the individual values are higher and some of them are lower, thus indicating that the basin is somewhat distorted by faults and joints. Again, the graph depicting the relation of number of streams of each order to order number shows a marked concavity at the lower end. If all the points were joined, as has been done by Horton, it indicates that geometric progression is not closely observed in the higher orders. We can further conclude that with the increase in order the number of stream decreases and this upholds the Horton's law.

By studing stream frequency of the basins a perfect interrelationship between stream order and drainage texture has been established. The values of stream frequency per sq km are high within the lower stream order which indicates the zone of fine texture. The low frequency values are found within higher order. As the stream order becomes higher, the value of frequency also becomes higher, which indicates the approaching maturity of the basin development. The higher stream frequency value in case of Ihang and Ithoi indicates a higher and more dissected terrain than other basins. Wari Lok has small basin area but the stream frequency is much higher in the first order and less mature in the advance stage, thus indicating higher scarp with dissected terrain. Leimakhong shows the lowest value of stream frequency, thus indicating mature and advance development of the basin in the higher stream order. Among all these basins, Leimakhong is the most matured drainage basin followed by Ekou Lok.

The study of length of stream with its mean values in relation to order number are of great importance while analysing the drainage characteristics of a basin. As the stream order goes higher the length of stream also becomes higher. The cumulative mean length also shows the same pattern showing progressive increase with increase in order. The lower value of mean length for the first order streams shows that most of them are short and, therefore, indicate more dissected areas.

As estimated by Horton, the length ratio should range between 2 and 3. However, the length ratio of the drainage basin of Iril river do not totally conform to that of Horton. The number of streams is dependent on the bifurcation ratio and increases with stream order while the length per stream is dependent on the stream length ratio. Further, according to Horton, mean drainage area should form a geometric series. The mean stream length and the mean areas of the drainage basins or each order have a geometric relationship between them. In this catchment all the selected drainage basins show close relationship between the stream order and the average drainage basin area.

High drainage density shows high relief region with higher amount of rainfall. Wari basin shows the highest drainage density followed by Ithoi and Ihang. Leimakhong shows the lowest drainage density, thus showing a less dissected and lower relief region. The Ekou and Ichai basins are lower than Wari, Ihang and Ithoi, thus indicating them to be of somewhat lower gradient than the other three drainage basins.

For determining various characteristics such as hydrological regime, physiographical character and erodibility of a drainage basin, length of overland flow is one of the important independent variables. Leimakhong drainage basin shows the highest length of overland flow, which may be due to the dominant valley portion in its basin area. Due to steeper slopes and highly undulating nature of the landscape, the length of overland flow is much lower in the basins of Wari, Ithoi, Ihang, Ekou and Ichai. For observing the nature of run off, surface erosion and drainage network development, the measurement of the constant of channel maintenance is highly required. The Wari, Ihang and Ithoi sub-basins require the minimum area for the development of drainage channel whereas the Leimakhong sub - basin needs the maximum area for the development of its drainage channel.

The morphology of an area is expressed by the appearance, dimension and magnitude of its slopes. Regional variation in slopes is associated with tectonic movement with high degree of denudational activity besides other factors such as weathering, process of erosion, transportation, etc. which have shaped the slope of the area. Average relief in different categories express the intensity of relief, which can be determined per unit area. In the Iril basin the highest intensity of relief covers 9.73 per cent, moderate intensity 55.25 per cent and the lowest intensity 35.02 per cent of the entire basin. The area of high intensity is confined to the northern part and moderate intensity almost scattered around the central portion of the basin. The low intensity of relief covers the lower middle portion of the basin and extends upto the mouth part.

The relative relief map of the area expresses the amount of slope and nature of the surface slope. The surface with relative relief ranging from 300 m to 500 m and above covers 23.08 per cent of the total basin area mostly in northern and north eastern parts, showing a highly dissected and undulating topography. Moderate relative relief, between 100 m and 300 m, cover most parts of the basin, *i.e.* 61.51 per cent of the total basin area. It is scattered in the western most part, central and southern parts of the Iril basin. Mapao Tangkhul and Yasmongjong villages and the southern portion of the basin have moderate relative relief. The lowest relative relief region occupies 15.41 per cent of the total basin area in the southern part around Ihang Karong, Sagolmang, Ekou Bazar and Pukhao area, besides the entire plain sector. Absolute relief can reveal the direct or indirect evidences of the genesis and evolution of the landforms.

The area of absolute relief having the height range of 1000 m - 1200 m has the largest coverage in the Iril basin. Villages like Shongben, Phaikon, Matakhong, Mollen, Sanakeithel Bunglung Makeng, and Saikul, etc. are situated within this relief region which covers 38.69 per cent of the total basin area.

The lowest relief region with elevation below 1000 m covers

most of the southern part. It has a flat plain terrain and covers 23.01 per cent of the total basin area. The villages Sagolmang, Pukhao, Lamboikhul, Taretkhul, Nongda, Kanglasiphai, Naharup, Irilbung, Urup Arapti and Lilong are all situated within this height range. The height range of 1200 m - 1400 m covers 17.21 per cent of the basin. It is associated with high hills which have the least of human settlements. Of course, some of the villages such as Washangphung Naga, Nurathel, Mongbung, Shorbung Naga, New Ngaton and Old Ngaton, etc. are situated in this region. The region below 1200 m - 1600 m defined as medium high land area, also has some settlements, but the highest altitude zone, with above 2200 m range, has the least area covering 0.41 per cent of the basin. It is associated with high escarpments, dense forests and lack of settlements.

Dissection index of this basin can give an idea for estimating the vertical balance of erosion. The low dissection index covers 15.67 per cent of the total area and is associated with low relief region. The medium dissection index is dominant in the northern and north eastern parts covering 55.46 per cent of the basin area. About 28.67 per cent of the total basin area has high dissection index and is associated with high relief. A comparison of dissection index map with relative relief map also supports the above facts and, therefore, it can be stated that the region above 300 m is well dissected, the region between 100 m and 300 m moderately dissected and the region below 100 m less dissected.

The generalised contour map helps in reconstructing the past features of the land surface. A comparison of the past (reconstructed contour) and present maps reveal that high surfaces in the past have been destroyed by various agents of weathering, erosion, etc. and, therefore, the surfaces which show steep faces today might have been less steeper in the past. The area in between 800 m and 1000 m seems to have gained a lot of land as compared with the past. This may be due to continued destruction by the tributaries of Iril river as well as other agents. The higher surface areas have been worn down, transported and deposited in the lower relief areas. In comparison with the high and lower relief surfaces, the lower surfaces have faced the least destruction and received more areas from the higher dissected and eroded parts. The study of stream line surfaces indicates the amount of denudation yet to be achieved. By drawing stream line surfaces, it is seen that the eastern and southern highland parts of the basin area are liable to denudation. The medium high and lowest stream line surfaces cover extensively the middle and southern parts of the basin which is an indication of land available for denudation.

By analysing the average slope the study area has been divided into three slope regions :

a.	Gentle slope area	8^0 and below
b.	Moderate to moderately steep slope area	8^0 - 20^0
c.	Steep slope to very steep slope area	20^0 and above

The gentle slope area covers 6.02 per cent of the total basin area and coincides with low relief region. The moderate to moderately steep slope area covers most of the southern half of the basin forming around 29.73 per cent of the entire basin. The steep to very steep slope area covers 55.41 per cent of the Iril basin comprising mostly hilly and highly dissected tracts.

The study of hypsometric curve and hypsometric integral value reveals the stage reached by a drainage basin and the percentage of mass removed from the basin by different erosional processes. Iril river shows the hypsometric integral value of 37.77 per cent which is little above the rough limit of 35 per cent set by Strahler. However, the overall characteristic features confirm its senility and monadnock phase.

An analysis of area height curve shows that the land between 1900 m and 1600 m presents a sharp diminishing trend, thus indicating intensive erosion and deep dissection. The longitudinal profile of a number of drainages in the Iril basin reveals the break in slope and phases of their life cycles. The Iril river shows three phases in its life cycle during her journey from the source to the mouth. As most of the streams start from the high relief, they grade downward suddenly and attain the balance between erosion and deposition of a matured river within a short span (of journey).

The cross profile of the Iril landscape shows the stages in the cycle of erosion, breaks in slope, terrace formation, accordant of summit level, etc.. Here, six accordants of summit level of 1800 m, 1600 m, 1400 m, 1280 m, 1200 m and 1000 m are distinguished. The study of accordants of summit level shows erosional surfaces at the levels of 1000 m and 900 m and general relief towards NW - SE direction. The undulating hills also show that the maximum part of the basin is dissected. In addition to the cross profile of the landscape, cross or traverse valley profiles have also been drawn and the stage of maturity gained by the Iril river in its course of journey has been visualized.

An attempt has been made to classify landforms on the basis of comparison of the results of different morphological techniques applied to the different tributaries of Iril river and their basins. Five types of landform have been recognised such as :

(a) High hills

(b) Dissected hills

(i) Highly dissected hills.
(ii) Medium high dissected hills.

(c) Lower valley region intermixed with residual hills, and

(d) Lower plain region.

Out of 1285 sq km area of the Iril basin, degraded forests cover 51.53 per cent, current *jhum* 32.21 per cent, cultivated land in the valley 14.53 per cent, forest blank 1.67 per cent and the built up land 0.06 per cent. The vast extent of scrub forests have been resulted from widespread *jhuming,* wildfires and indiscriminate deforestation, etc. The lands under current *jhum* are very unproductive and environmentally unstable. The cultivated lands in the valley area are also monsoon fed and in the absense of irrigation facilities these precious lands remain fallow during winter. The too limited extent of built up land indicates less urbanisation and for that matter the backwardness of the area.

Based on the major findings of the study, landuse planning and watershed management of Iril river basin have been attempted and portrayed through maps related to potential areas for land development and spatial planning. Under an integrated approach, various suggestions have been made for treatment of degraded forests, agricultural lands and wastelands, etc. Subsidiary occupations for the local inhabitants have been identified for their economic rehabilitation.

Large scale afforestation of the hills with timber species in upper reaches and timber, firewood and fodder species in the height range 900 - 1800m has been suggested. This must go concomitant with the practice of selective logging by banning the indiscriminate tree felling operation by the contractors.

For treatment of the degraded *jhum*lands toposequencing of crops as forest - silvipastoral horticultural - terrace cultivation downhill with various soil and water conservation measures like contour bunding, terracing, check dams, water storage cisterns, etc. have been suggested on the hill slopes between 7^0 and 17^0.

For enhancing the yield of the wet agricultural fields in the slope range 0 - 6^0, certain measures like minor irrigation, scientific rotation of crops and modern inputs like HYV (High Yielding Variety) seeds, fertilizers and pesticides have been suggested.

Reclamation of different kinds of wastelands as degraded hillslopes, undeveloped piedmont areas and the derelict wetland of Phaknung *Pat* has been proposed with the identification of appropriate landuses. These wastelands can be put to pasture development with scientific range management strategy. It has been suggested for combating the harmful effect of grazing.

Sustainable development of tribal economy can be achieved through subsidiary occupations. Therefore, livestock farming, piggery, poultry, mushroom cultivation, pisciculture and beekeeping along with various agro - based small scale industries like dairying, fruit processing, cane and bamboo works, dehydration and canning plants and rice mills have been proposed at certain vantage points.

An intensified community development programme with special components like adequate food storage system, better communication, efficient biomass energy exploitation devices, improvised agricultural and farming tools, literacy drive, environmental education, etc. have been proposed for the overall economic prosperity of the people in the river basin.

Iril river basin can be a model watershed for adoption of scientific management for environmentally sound economic development.

BIBLIOGRAPHY

Agrawal, M. (1972), Morphometric analysis of some drainage basins of Jabalpur plateau, *National Geographical Journal of India*, Vol. 18 (3 - 4), pp. 148 - 168.

Ansari, S.A. (1976), *Economic Geography of Manipur*, Trio Book Store, Imphal.

Arogyaswamy, R.N.P. (1967), Geomorphological features of the Nilgiri mountains, *Proc. Geom. Stud. Ind.*, pp. 78 - 88.

Arunachalam, B. (1972), Some salient geomorphic features of south Ratnagiri region, *Proceedings of Symposium on Geomorphology of Indian Shield*, 21st International Geographical Congress, India, 1968, National Committee for Geography, Calcutta, pp.27 - 32.

Auden, J.B. (1954), Erosional pattern and fracture zones in Peninsular India, *Geol. Magazine*, Vol. 140 (2), pp. 89 - 101.

Bagchi, K. (1960), Drainage pattern in Ganga delta, *Geog. Rev. of India*, Vol. 22 (4), pp. 51 - 53.

Ball, V. (1881), The geology of Manbhum and Singbhum, *Mem. G.S.I.*, Vol. 18 (2).

Bandhopadhyay, K.A. (1974), Mechanism of rills: An investigation on microgeomorphology, *Geo. Rev. of Ind.*, Vol. 35 (3), pp. 264 - 275.

Basu, S. (1976), Some fundamental problems of meander formation with special reference to the Bhagirathi river, *Geo. Rev. of Ind.*, Vol. 38 (1), pp. 54 - 67.

Basū, S. and Ranga Raju, K.M. (1964), Geology of the Manipur valley and adjacent hills, Report of the ONGC (unpublished).

Betal, H.R. (1973), Geomorphological mapping of Frajerganj coast, *Geo. Rev. of Ind.*, Vol. 35 (2), pp. 192 - 195.

Bhattacharya, R. and Das, p. (1973), A Geomorphological account of Kaliaghai river basin, *Geo. Rev. of Ind.*, Vol. 35 (4), pp. 365 - 375.

Chakraborty, S.C. and Ghosh, A.R. (1974), Longitudinal profiles in fluvial process : An analytical study, *Geo. Rev. of Ind.*, Vol. 36 (1), pp. 23 -37.

Chatterjee, S.C. (1945), Some aspects of the geomorphology of the Ranchi plateau, *Geo. Rev. of Ind.*, Vol. 7 (1) pp. 31 - 35.

Chatterjee, S.P. (1968), *Progress of Geography in India*, Indian Science Congress Association, Calcutta.

Chibber, H.L. (1953), The landforms life and radial drainage of Mount Parasnath, Hazaribagh district, *Nat. Geo. Soc. Ind. Bull.*, No.18.

Chorley, R.J. (1969), The drainage basin as the fundamental geomorphic unit, in Chorley, R.J. (Ed), *Water, Earth and Man*, Metheun and Co., London, pp. 77 - 99.

Choudhuri, G. (1985), Quantitative analysis of the erosion surfaces in the Garra basin, *Indian Journal of Landscape Systems and Ecological Studies*, Vol. 8 (1), pp. 54 - 59.

Clark, J.I. (1966), Morphometry from maps, in Durry, G.H. (Ed), *Essay in Geomorphology*, Heinemann, London, p. 241.

Das, P. (1975), An appraisal of flood incidence of the Kaliaghai river basin, *Geo. Rev. of Ind.*, Vol. 37 (1), pp. 73 - 80.

Davis, W.M. (1915), Biographical memoir of John Wesley Powell, *Nat. Acad. Sci., Mem.*, 8, pp. 11- 83.

Devis, W.M. (1954), The rivers and valleys of Pennsylvania, Geog. *Essays*, Dover, p.261.

Dayal, B. and Duara, B.K. (1963), Report on the systematic mapping and mineral survey in parts of Manipur state, Progress report for Field Season 1962 - 63, Geological Survey of India (Unpublished).

Dikshit, K.R. (1976), Drainage basin of Konkan : forms and characteristics, *Nat. Geo. Journ. of India,* Vol. 22 (1), P.79.

Dunn, J.A. (1929), The geology of north Singbhum including parts of Ranchi and Manbhum districts, *Mem, Geological Survey of India,* Vol. 54, p.132.

Dury, D.H. (1952), Methods of cartographical analysis in geomorphological research, *Ind. Geo. Soc., Silver Jubilee Souvenir*, p. 139.

Dury, G.H. (1951), Quantitative measurement of available relief and of depth dissection, *Geol. Mag.,* Vol.28, p. 343.

Dury, G.H. (1969), Perspectives on geomorphic processes, *Assoc. American Geogr. Resources Pub.* No. 3.

Eric, A.C. (1967), Slope profiles of Mourne mountains, Dow, *Irish Geog.,* Vol.5, pp. 311 - 318.

Evans, P. (1932), Explanatory notes to accompany a table showing the Tertiary succession in Assam, *Trans. Min. and Geol. Inst. of India.* Vol. 27.

Evans, P.(1964), The tectonic framework of Assam, *Journ. Geol. Soc. India,* Vol. 5, pp. 80 - 96.

Gardiner, V. (1986), Fluvial paleohydrology : The morphometric contribution, *National Geographer,* Vol. 21(1), pp. 17 - 31.

Goudie, A. *et al* (1990), *Geomorphological Techniques,* Unwin Hyman, London.

Gour, M.P. and Khan, I.K. (1984), Report on the systemetic geological mapping around Paoyi Paorei, Tolloi Huimi area of Manipur East district, Manipur, progress report for Field Survey 1984 - 1985, Geological Survey of India (Unpublished).

Greenwood, G. (1857), *Rain and Rivers on Hutton and Playfair against Lyell and all comers,* 1st ed., London.

Gregory, K. J. (1977), *River Channel Changes,* Wiley, Chichester.

Goswami D. C., Dutta P. C. and Kalita, N.R. (1991), Braiding of the Brahmaputra river channel in Assam : A fluvio - geomorphological enquiry, *North - Eastern Geographer,* Vol. 23 (1 and 2), pp. 52 - 60.

Hettner, A. (1928), *Surface features of the land*, translated by Phillip Tilly (1972), The Macmillan Press, London, Vol. 29, pp. 14 - 15.

Horton, R.E. (1932), Drainage Basin Characteristics, *Trans. Am. Geophys.* Union, Vol 14, pp. 350 - 61.

Horton, R.E. (1945), Erosional development of streams and their drainage basins : hydrophysical approach to quantitative morphology, *Bull. Geol. Soc. Am.*, No. 56, pp. 275 - 370.

Hutton, J. (1975), *Theory of the Earth*, Vol. 1 and 2, Edinburgh.

Jagadeesh, J. and Karennavar, M.F. (1985), Quantitative analysis of streams in Varada river basin, *Ind. Journ. of Landscape Systems and Ecological Studies*, Vol. 8(1), pp. 72 - 78.

Joshi, R.V., Rajguru, B.N., Pappu, R.S., Thokchom, A.S.and Singh, O.K. (1987), Explorations in Manipur (Unpublished).

Jukes, J.B. (1862), On the mode of formation of some of the river valleys in the south of Ireland, *Q.J.G.S*, Vol., 18, pp. 378 - 403.

Khan, I.K., Jayaraman, V. and Mishra, V.K. (1985), Report on systematic geological mapping and palaeontological aspects around Saikul (Senapati district), Sanakeithel, Mulam and Tuinem (Ukhrul district) and around Sadu (Imphal district) of Manipur State, Progress report for Field Season 1984 - 1985, Geological Survey of India (Unpublished).

Khan I.K., Sabale, A.B. and Mishra, U.K. (1986), Systematic geological mapping and paleontological aspect of the area around Thawai Tangkul, Chadung Tangkul (Ukhrul district), Island Thaiyong (Senapati district) and around Lamlai, Seijang (Imphal district) of Manipur state, Progress report for the Field Season, 1985 - 1986, Geological Survey of India (Unpublished).

Khan, S.A. and Rao, V.K. (1989). Geology and structure of the areas around Sagolmang, Yaingangpokpi (Imphal district), Litan, Itham Tangkhul and Maphitel range (Ukhrul district) Manipur state, Progress report for Field Season 1987 - 1988, Geological Survey of India (Unpublished).

Kharkwal, S.C. (1968) Classification of Kumaon Himalaya into morpho-units, *Nat. Geog. Journ. of India,* Vol. 14(4), pp. 223-238.

Kharkwal, S.C. (1970), Morphometric study of a Himalayan basin : A sample study, *Nat. Geog. Journ. of India,* Vol. 16(1), pp. 47 - 60.

King, C.A.M. (1966), *Techniques in Geomorphology,* Edward Arnold, London.

King C.A.M. (1976), Landforms and geomorphology, *Bench Mark Papers in Geology,* Vol. 28, pp. 136 - 137.

Krishnan, M.S. (1953), The structural and tectonic history of India, *Mem. GSI,* Vol. 81, pp. 9.

Kumar, A. (1971), Geomorphic interpretation of drainage and relief of eastern Ranchi plateau, *Uttar Bharat Bhoogal Patrika,* Vol. 7(1), pp. 18 - 33.

Kumar, A. (1979), *Geomorphology of Simdega and its adjoining area, Bihar,* The Nat. Geog. Soc. of India, Varanasi.

Kumar, A. (1976), Morphological properties of the slope profiles of the western plateau scarp, Ranchi, *Geog. Outlook,* Vol. 11, pp. 83 - 94.

Kumar, A. and Pandey, R.N. (1977), Numerical analysis of small drainage basins of Hazaribagh plateau region, *Geog. Bull. Ind.,* Vol.2, pp. 1 - 13.

Kumar, A. and Singh, N.P. (1978), Quantitative classification of landform regions of the western higher plateau, Bihar, *Geog. Rev. of India,* Vol. 40(4), pp. 378 - 385.

Kushwaha, R.A.S. (1986), Geomorphological studies along Beas River Basin, Ph.D. thesis, Delhi University (Unpublished).

Leopold, L.B., Wolman, M.G. and Miller, J.P. (1964), *Fluvial processes in geomorphology,* Eurasia Publishing house, New Delhi, pp. 134 - 138.

Lubowe, J.K. (1964), Stream junction angles in the dendritic drainage pattern, *Am. Journ.* Sc. Vol. 262, pp. 337.

Lyell, S.C. (1872), *Principles of Geology,* Appleton and Co. New York.

Mache, R.N. and Peshwa, V.V. (1978), Photo - geological interpretation of the control on drainage in Gondwana and Bijawars of Son valley, Shahdol district, M.P., *Proceedings of Symposium on Morphology and Evolution of Landforms,* University of Delhi, pp. 250 - 254.

Mallet, F.R. (1876), On the coalfields of Naga hills bordering the Lakhimpur and Sibsagar district, Assam, *Mem., GSI,* Vol. 12, pp. 286.

Matley, I.M. (1966), The Marxist approach to the geographical environment, *Ann. Assn. American Geo.* Vol 56, pp. 97 - 111.

Miller A. (1953), *The Skin of the Earth,* Methuen and Co. Ltd., London, pp. 75.

Monkhouse, F.I. and Wilkinson, H.R. (1980), *Maps and Diagrams,* B.I.publication, New Delhi.

Mukherjee, A.B. (1975), Geomorphic study of Choe terraces of the Chandigarh Siwalik hills, India, *Himalayan Geology,* Vol. 5, pp. 302 - 306.

Mukhopadhyay, S.C. (1973), River terraces of the Subarnarekha basin, *Geog. Rev. of India,* Vol. 39(3), pp. 248 - 258.

Mukhopadhyay, S.C. (1980), *Geomorphology of the Subarnarekha basin,* The University of Burdwan, pp. 117.

Mukhopadhyay, S.C. (1988), Some aspects of geomorphology of Manipur area, *Indian Journ. of Landscape System and Ecological Studies,* Vol. 11 (2), pp. 182 - 198.

NIRD (1989), *Courses on Watershed management : Reading Material,* National Institute of Rural Development Hyderabad, pp. 1-94.

Oldham, R.D. (1883), Report on the geology of Manipur state and Naga hills, *Mem. Geological Survey of India,* Vol. 19 (4).

Padmaja, G.(1975), Some aspects of quantitative drainage characteristics of the Dhund basin, *Geo. Rev. of India,* Vol. 38(2), pp. 158 - 164.

Pal, S.K. (1973), Quantitative Geomorphology of drainage basins in the Himalaya, *Geo. Rev. of India*, Vol. 35(1). pp. 81 - 101.

Pal, S.K. and Bagchi, K.(1975), Recurrance of floods in Brahmaputra and Kosi basins : A study in climatic geomorphology, *Geo. Rev. of India*, Vol. 37(3), pp. 242 - 248.

Pandey, R.K. (1985), Morphometric personality of Bhilangana network (U.P.Himalaya), *Indian Journ. of Landscape System and Ecological Studies*, Vol. 8(1), pp. 34 - 40.

Pathak, M. (1967), A study of landforms in Jethian region, Gaya district Bihar, *Proceedings of Seminar on Geomorphological Studies in India, University of Sagar*, pp. 49 - 55.

Pascoe, E.H. (1912), Traverse across Naga hills of Assam from Dimapur to the neighbourhood of Saramati peak, *Records of the Geological Survey of India*. Vol. 42(4).

Penck, W. (1953), *Morphological Analysis of Landforms*, (First edition 1924), Macmillan and Co., London.

Powell, J.W.(1875), *Explanation of the Colorado river of the west*, Govt. Printing press, Washington.

Prahlad, K.(1986), Morphology of drainage lines : A study of a greater Himalaya basin (the Dhauli Ganga), *Nat. Geo. Journ. of India*, Vol. 32(3), pp. 235 - 245.

Radhakrishnan, B.P. (1952), The Mysore plateau, its structural and physiographical evolution, *Bull. Mysore Geologists Assn.* Vol. 3, pp. 57.

Radhakrishnan, B.P. (1964), Evolution of the Sharavati drainage Mysore state, S.India, *Journ. of the Geol. Soc. of India*, Vol. 5, pp. 72.

Raghunath, H.M. (1986), *Hydrology : Principles, analysis and design*, Wiley Eastern Ltd. New Delhi, pp. 97 - 120.

Raisz, E. and Henry, J. (1937), An average slope map of southern New England, *Geo. Rev.*, Vol. 27, pp. 417 - 472.

Ray, R. and Chanda, S.(1989), Late Quaternary vegetational history and datings of some sediments of Loktak Lake of Manipur, *Journ. of Palynology*, Vol. 25 (1).

Richards, K.S. (1981), Introduction to morphometry, in Andrew Gaudie *et al* (Ed), *Geomorphological Techniques*, Georg Allen and Unwin, London, pp. 25 - 26.

Richards, K.S. (1982), *Rivers : Forms and process in Alluvial Channels*, Methuen, London.

Sabale, A.B. (1988), Report on systematic geological mapping of the area around Tentha Khunou and Heirok villages of Thoubal district and Machi village of Chandel district, Manipur State, Progress report for the Field Season, 1987 - 88, Geological survey of India (Unpublished).

Satpathi, D.P.P. (1972), Quantitative analysis of landform : A case study in the Deo river basin, *Geographical Outlook*, Vol. 9, pp. 57 - 66.

Satpathi, D.P.P. (1976), Meander mechanism, geometry of meanders and sinuosity index of the major river of Singbhum, India, *Geographical Outlook*, Vol. 12, pp. 11 - 22.

Satpathi, D.P.P. (1981), *An outlook of Indian Geomorphology*, Classical Publishing Co., New Delhi.

Schumm, S.A. (1956), Evolution of drainage systems and slope in badlands at Perth Amboy, New Jersey, *Bull, Geo. Soc. America*, Vol. 63.

Selby, M.J. (1982), *Hillslope Material and Processes*, Oxford University Press, Oxford.

Sen, R.N. (1977), Recent aspects of the geomorphic evolution of Bandi river basin, E. Rajasthan, *Geo. Rev. Of India*, Vol. 34 (3), pp. 353 - 366.

Sharma, H.S. and Padmaja, G.(1977), Quantitative geomorphic characteristic of stream of the Moral basin, Rajasthan, *Geo. Rev. of India*, Vol. 39(4), pp. 353 - 366.

Shukla, R.C. and Mahapatra, N.R. (1988), Geoenvironmental evolution of Imphal valley including the Loktak lake and its catchment areas, Manipur, Progress report for the Field Season 1985 - 86, Geological Survey of India (Unpublished).

Shukla, R.C., Mahapatra, N.R. and Hedge, S.V. (1989), Quaternary geology and geomorphology of a part of Imphal valley, Manipur central district, Manipur, Progress report for the Field season. 1984 - 85, Geological Survey of India (Unpublished).

Sidhu, G.S. (1974), Parametric analysis of Patalgad drainage basin east of Mussoorie, *Nat. Geog. Journ. of India*, Vol. 20(4).

Singh, H.T. and Shyamananda, R.K. (1988), Status Report on the Environment of Manipur, Dept. of Science and Technology, Govt. of Manipur, Imphal (Unpublished).

Singh, J. (1979), *Central places and spatial organisation in a backward economy : Gorakhpur region - A study in integrated regional development*, Uttar Bharat Bhoogol Parishad, Gorakhpur.

Singh, N.P. (1975), Geomorphology of the western higher plateau of Bihar, Ph.D. thesis, Magadh University, Bodhagaya (Unpublished).

Singh, R.P. (1958), Structure, drainage and morphometry of the Chotanagpur highland, *Geographical Outlook*, Vol. 1, pp. 23 - 32.

Singh, R.P.(1969), *Geomorphological evolution of Chotanagpur Highland*, India, Nat. Geo. Soc. of India, Varanasi, pp. 29.

Singh, R.P. (1972), Geomorphological studies, in ICSSR (sponsored) *A survey of research in physical geography*, Concept Publishing Company, New Delhi, pp. 7 - 46.

Singh, R.P. (1982), *Geography of Manipur*, National Book Trust, India, New Delhi.

Singh, S. (1974), The cycle of erosion : Its changing perspective, *National Geographer*, Vol. 9, pp. 99 - 101.

Singh, S. (1981), Estimation of drainage density, *National Geographer*, Vol. 16 (2), pp. 81 - 98.

Singh, U. (1986), Morphological characteristics of some selected river basins of the middle Narmada valley, *Nat. Geo. Journ. of India*, Vol. 32 (3), pp. 261 - 274.

Singh, V.R. (1970), *Land use Pattern in Mirzapur and Environs*, Banaras Hindu University, Varanasi, pp. 111.

Smith, G.H. (1935), The relative relief of Ohio, *Geographical Review*, Vol. 25, pp. 272 - 284.

Srivastava, S.N.P., Majumdar, M., Chattopadhyay, B.and Roy, R.K. (1976), Mineralised ultramafic belt of Manipur and Nagaland, India, *Proceedings of 17th Himalayan Geology Seminar*, 1976.

Stocking, M.A. (1972), Relief analysis and soil erosion in Rhodesia using multivariate techniques *ztschr, fgeo*, Vol. 16 (4), pp. 432 - 433.

Strahler, A.N. (1950), Equilibrium theory of erosional slopes approached by frequency distribution analysis, *Am. Journ. Sc.*, Vol. 248, pp. 673 - 96.

Strahler, A.N. (1952), Dynamic basis of geomorphology, *Bull., Geol. Soc. Am.*, Vol. 63, pp. 923 - 38.

Strahler, A.N. (1957), Quantitative analysis of watershed geomorphology, *Trans. Am. Geophys. Union*, Vol. 38, pp. 913 - 20.

Strahler, A.N. (1964), Quantitative geomorphology of drainage basins and channel networks, in Chow, V.T. (ed) *Handbook of Applied Hydrology*, Mc Graw Hill, New York, pp. 411.

Strahler, A.N. (1971), *Physical Geography*, Wiley Eastern, New Delhi.

Theobold, W. (1873), On the geology of Pegu, *Mem. Geological Survey of India*, Vol. 101.

Thornbury, W.D. (1954), *Principles of geomorphology*, John Wiley and Sons, New York, pp. 5 - 14.

Vaidyanathan, R. (1977), Recent advances in Geomorphic studies of Peninsular India : A Review, *Ind. J. Earth Sc.*

Verma, V.K. and Tandon, S.K. (1971), Stream tortuosity, A new method of visual estimation, *Annals of Geomorphology*, N.F. Supplement, Berlin. Vol. 12, pp. 165 - 171.

Verma V.K. and Shukla, M.K. (1973), Morphostatistical investigations on drainage network around Pachmarhi (M.P.), *Recent researches in Geology*, Vol. 1, pp. 137 - 145.

Verstappen, H.T. (1983), *Applied Geomorphology*, Elsevier, pp. 10 - 11.

Wadia, D.N. (1937), An outlook of geological history of India, in *An outline of the field science of India*, Ind. Sc. Congress Assn., Calcutta, pp. 43 - 69.

Wentworth, C.K. (1930), A simplified method of determining the average slope of land surface, *Am. Journ. of Sc.*, Vol. 20, pp. 184 - 194.

INDEX

Plates

3.1 A Section of the Iril river bank

3.2 Deposition of eroded materials near Naharup

5.1 Confluence of Iril and Imphal rivers

5.2 Ichai Lok terraces used for cultivation

5.3 Ihang Lok terraces used for settlement

5.4 Channel bar at the confluence of Ithoi and Iril rivers

5.5 Meandering river course near Taretkhul

6.1 Deforested hill near Ekon Lok

6.2 Bare hill face near Saikul

6.3 A deforested hill affected by burning and felling of trees

6.4 Jhuming and wet cultivation near Saikul

6.5 Land use along river bank during winter season

6.6 Bamboo plantation along Iril river bank